HARLEY-DAVIDSON

A Way of Life

Sandcastle
Books

2-3 The 2010 Harley-Davidson CVO Street Glide FLHXSE; a special limited production of 3500 units of the famous "bagger" Electra Glide.

HARLEY-DAVIDSON
110

CONTENTS

Text and Photography
Albert Saladini
Pascal Szymezak

Graphic Design
Patrizia Balocco Lovisetti

Editorial coordination
Laura Accomazzo

Translation
Studio Traduzioni Vecchia, Milan
Hugh Swainston - Karla Donnelly

Sandcastle Books

This edition published 2014 by Sandcastle Books Ltd, Suite 3 The Workshop Business Centre, Pinvin, Worcestershire, WR10 2ES

ISBN 978-184999-274-9

1 2 3 4 5 6 18 17 16 15 14

Printed in China

4-5 The 2010 Harley-Davidson VRSCAW V-Rod, like the other models of the same successful series is powered by the Revolution engine.

Ride Free

PREFACE

Harley-Davidson occupies a unique position in global motorcycle culture. It is at once known as the purveyor of all that is American, and for being an international symbol of freedom. It is universally recognized as an institution for those who ride free.

Albert Saladini understands this marvelous position occupied by the Harley-Davidson Motor Company, and knows that the mystique is based entirely upon the motorcycles themselves: he is a rider, scholar, photographer, and brother.

Above all, though, he is a Harley-Davidson motorcyclist, and thus brings all this to bear upon what he feels in the wind upon a Harley. He subjects his scholarship and photography to the ultimate test, which is how it best signifies what it means to ride a Harley-Davidson motorcycle. For this reason, his work is not only accurate and aesthetically correct . . . it is righteous!

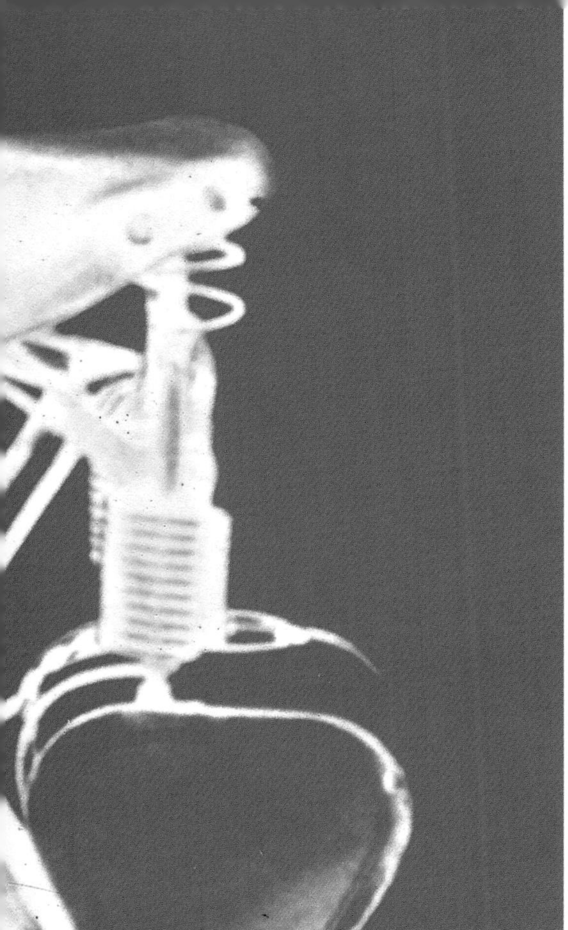

***7* Martin Jack Rosenblum, a Harley-Davidson historian, first became known as a poet and musician under the name of "Holy Ranger." After publishing a collection of poems and three CDs on the glory of the Harley-Davidson motorcycle, he now works to save and preserve its historical heritage. Harley-Davidson celebrated its one hundredth anniversary in 2003.**

8-9 The 2010 Harley-Davidson FLSTFB Softail is a new "darker" and lower version of the successful Fat Boy.

HARLEY
DAVIDSON
USA

HARLEY-DAVIDSON
HARLEY-DAVIDSON
HARLEY-
MOTOR
HARLEY-DAVIDSON
CYCLES

Introduction

She is a true "grande dame" who, in over 100 years, has never developed a wrinkle and has demonstrated again and again that classics never go out of style. The color of her dress doesn't matter; what is important is the way her rhinestones are displayed. She has had the greatest lovers, all passionate, who have covered hundreds of thousands of miles for her and with her. She has always been looked at, examined, undressed with the eyes, and envied. She has won great prizes and wars; she has gone places others could not. Today, she is still captivating; she is celebrated and exhibited, dressed to please different tastes. Her return has upset the market, which is flooded by strong Japanese competition.

She fears no one. She is more than 100 years old and, every morning when I wake her up and uncover her, I cannot help admiring her perfect line, which made my father dream and made me write this book.

Thank you, Mr. Harley. Thank you, the Davidsons.

Albert Saladini

Over the century of its rich and eventful history, the Harley-Davidson Motor Company has won the affection and loyalty of a vast range of customers. Its motorcycles have been used for day-to-day transport, for touring, for racing, and simply for fun; they have been ridden by soldiers in two world wars, by the hippies and other members of the culture in the 1960s, by cops, and by crooks; and they have served as status symbols for stars and starlets. But whatever the nationality, profession, or social status of a Harley owner, he or she shares one thing with all the others – a devotion to the products of the Harley-Davidson Motor Company. For whatever your dream may be, a Harley will make it come true, and it will bring you into contact with others who share the same dream. It is this unique magic which wins the affection of Harley riders and makes their hearts beat faster as soon as they feel the first throb of a Harley-Davidson engine.

My own Harley conjures up feelings to which only she has the key-sensations that re-awaken memories of wide-open spaces, of past journeys, of pure happiness. This book is intended as a tribute to this automotive goddess, whose history is as old as the century but whose charms remain as potent as ever; I offer it up, with gratitude, to a machine which has shared my most cherished moments of freedom and joy.

Pascal Szymezak

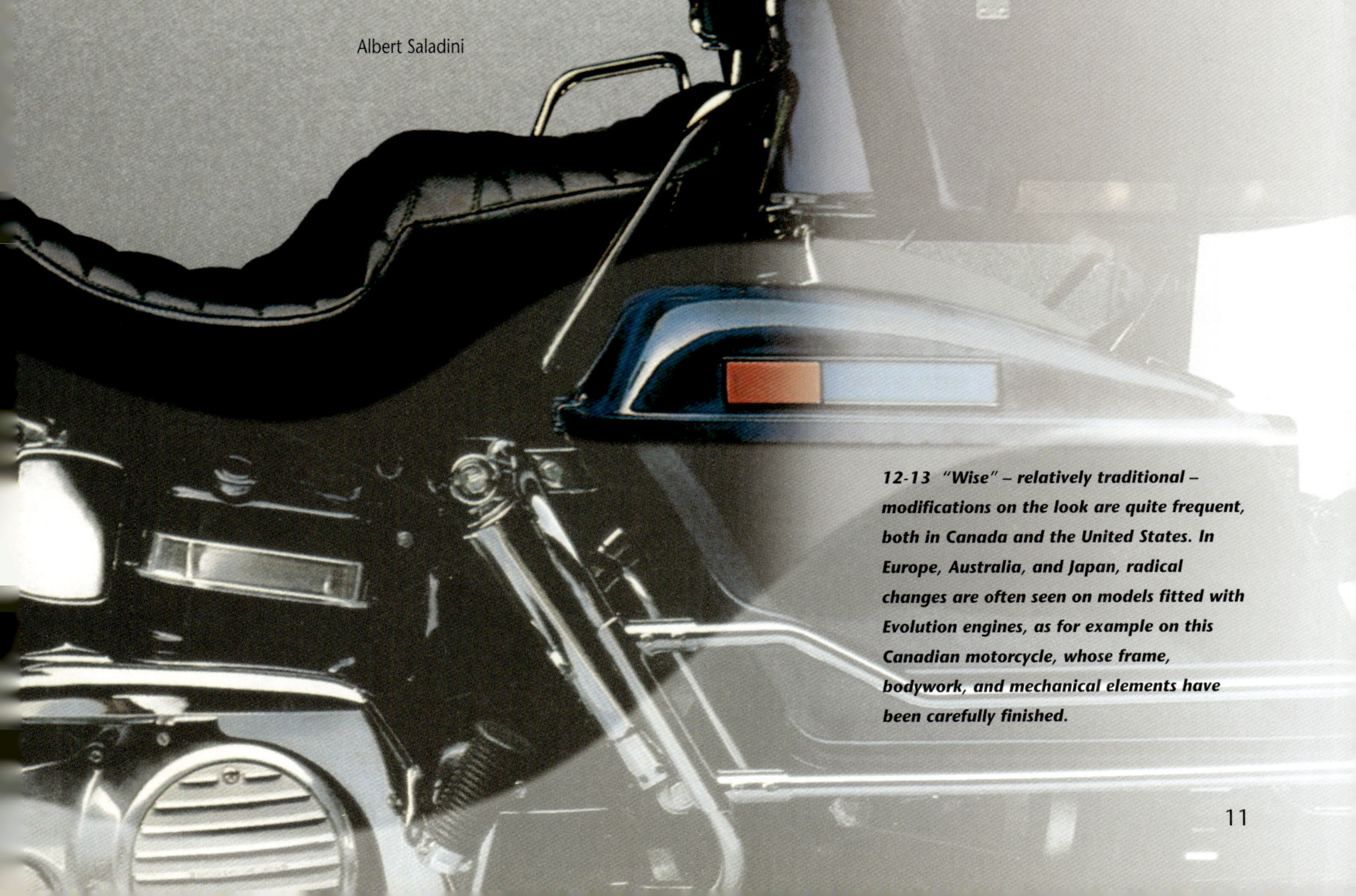

12-13 "Wise" – relatively traditional – modifications on the look are quite frequent, both in Canada and the United States. In Europe, Australia, and Japan, radical changes are often seen on models fitted with Evolution engines, as for example on this Canadian motorcycle, whose frame, bodywork, and mechanical elements have been carefully finished.

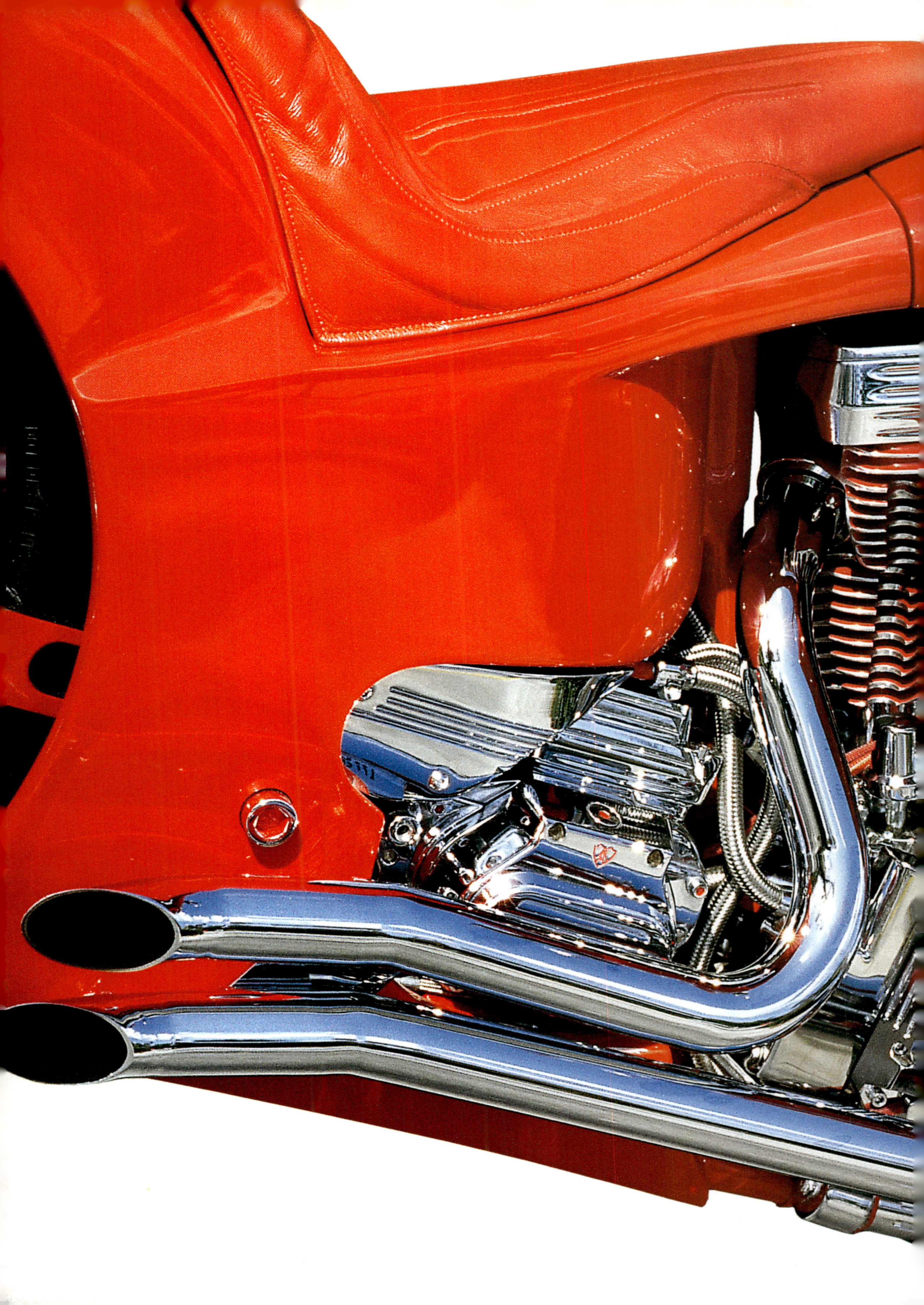

METAL HEART

Harley-Davidson: Past and Future

Not so long ago, there were those motorcyclists who believed that Harley-Davidsons were obsolete machines, relics from another era. They never took the time to examine or appreciate Harley-Davidsons, and few journalists made an effort to provide them with any information on the subject.

But in the last few years, there's been a growing enthusiasm about Harleys, along with an increase in the number of Harley owners. Companies that not so long ago disparaged the Milwaukee brand now manufacture motorcycles that imitate the Harley-Davidson look or that are, at the very least, strongly influenced by Harley-Davidson, with an engine architecture similar to that of the V-Twin. Riders are abandoning their ordinary sport bikes and heavyweight motorcycles for big Harley-Davidsons. Is the change some kind of turning point in the history of motorcycle manufacturing, or just a fashion decision? Either way, it underscores the sound thinking of the team at Harley-Davidson, who have relied on Harley-Davidson's heritage and its brand image to win over customers tired of the parade of novelties proposed by other manufacturers over the years. Today Harley-Davidson finds itself a leader in the world of custom and deluxe touring bikes, closely watched by its competitors, who strive to incorporate Harley concepts into their own machines.

A quick look at the history of Harley-Davidson shows that the company has always been ahead of the curve in utilizing state-of-the-art techniques. Back when production of Harleys began, the bikes were already fitted with twist grips – unlike European and Asian motorcycles, which only began offering riders twist grips in the middle of this century. In the 1920s, the first machine to be equipped with a side stand and

an electric lighting system (other brands still used acetylene headlights) was a Harley. And it was a Harley-Davidson that inaugurated the use of valve tappets with hydraulic return – even before the system was adopted for any mass-produced American car.

Another Harley-Davidson innovation, in 1981, was the adoption of a Kevlar fiber toothed belt secondary transmission on stepped pulleys. At the beginning of the century, the company had replaced the belt secondary transmission on its motorcycles with a chain transmission, while other manufacturers stuck with belts, which, at that time, skidded, broke, and required constant adjustment. Fittingly, Harley-Davidson was the first motorcycle company to go back to this early type of secondary transmission, which thanks to new, improved materials now has the advantage of requiring little maintenance. The tension needs to be readjusted only when the rear tire is changed, and the system can hold up for tens of thousands of miles – a big improvement over a chain secondary transmission, which not only makes the rear part of the bike dirty, but requires continuous maintenance and frequent replacement.

Apart from the fact that the Harley-Davidson engine offers an incomparable sensation of controlled power, formidable thrust even at low speed, considerable torque, and the feeling of a real living engine, the main thing going for Harley-Davidsons is their look. The Harley look changes somewhat from model to model but can be recognized by a long, raised handlebar, two half-tanks with the dashboard and speedometer located on top, a twin seat, plastic saddlebags, and a calm heavyweight motorcycle line – in short, all of those things that the people who now copy Harley-Davidson once mocked about this legendary brand. Harley Davidson, the oldest continuously operating motorcycle manufacturer in the world, in operation since 1903, has managed not only to preserve its most ingenious technical innovations over the years – the use of hydraulic tappets, for example, or toothed belt transmission – but to combine these innovations with new improvements, like the adoption, in 1984, of an engine made entirely of aluminum. Over time, Harley-Davidson has upgraded its techniques to keep up with scientific progress but has nevertheless maintained the Harley spirit and the Harley look.

Just talking about Harley-Davidsons isn't enough. You have to ride one of these motorcycles to fully understand what it offers to riders, what it lets them discover and share – to understand what it is to commune with a machine that will remain unique through the years and will give you renewed pleasure from the moment its wheels first turn each day.

Here is a motorcycle that doesn't look or act its age. Its very longevity is its charm.

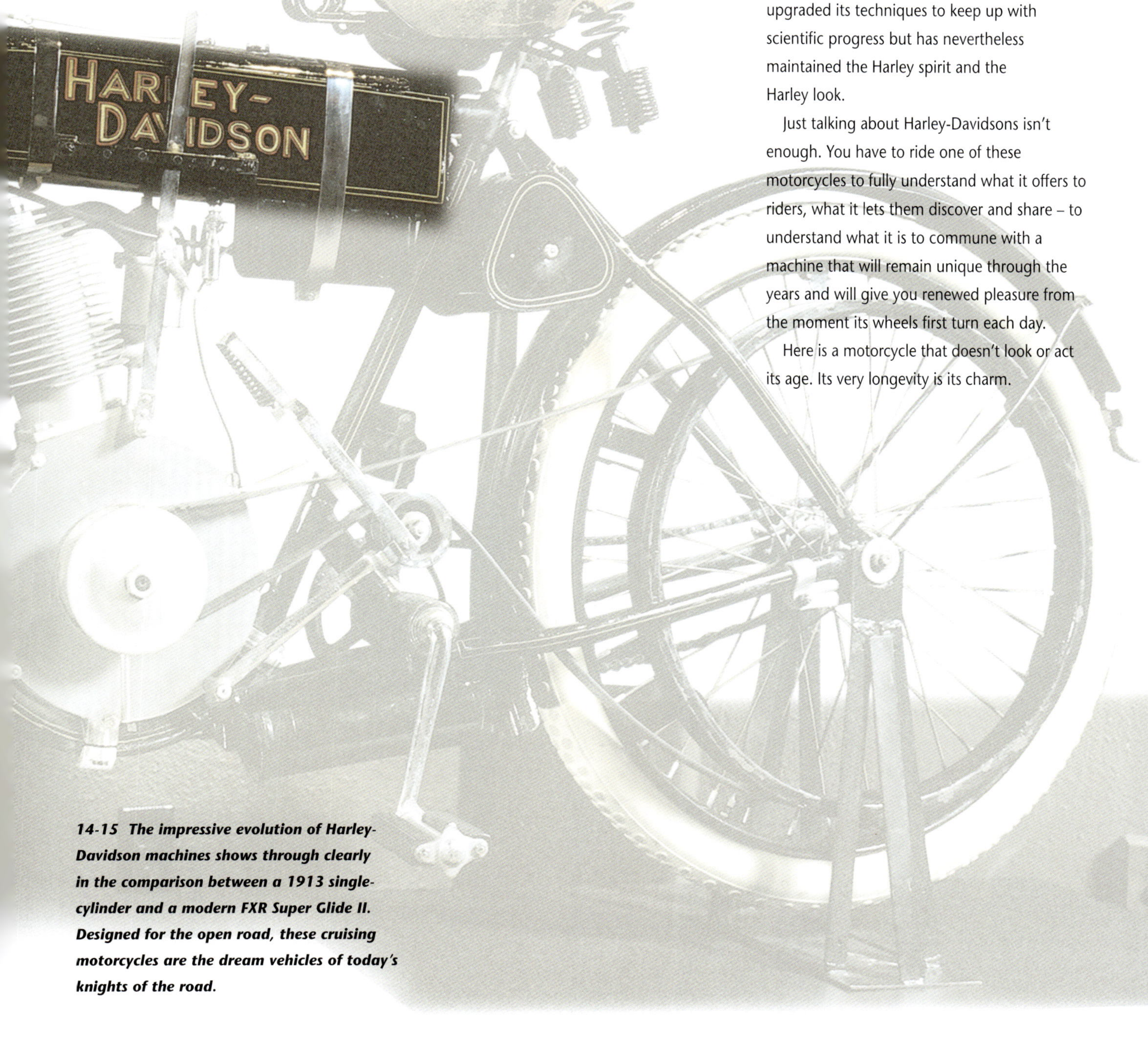

14-15 The impressive evolution of Harley-Davidson machines shows through clearly in the comparison between a 1913 single-cylinder and a modern FXR Super Glide II. Designed for the open road, these cruising motorcycles are the dream vehicles of today's knights of the road.

From Louis Guillaume Perreaux to Harley-Davidson

On Sunday, April 5, 1818, a vehicle called a draisienne, fitted with a back steam boiler and named "Velocipédraisiavaporiana," allegedly made its debut in the Luxembourg Gardens in Paris. The draisienne, a vehicle dating back to the end of the eighteenth century, was made entirely of wood (frame, wheels, and spokes) and looked like a bicycle without pedals, propelled by pushing with one's feet. Later, Pierre Michaux and his children had the idea of fitting some pedals to the draisienne's front wheel.

On December 24, 1869, L. G. Perreaux gave the Michaux family a small steam engine, which could be fitted to their "velocipedes" and, in 1870, the first "steam velocipede" was tested between Paris and Saint Germain en Laye over a distance of 16 kilometers.

In 1879, an Italian called Murnigotti patented a two-wheeler powered by a 0.5 horsepower four-stroke engine – but the machine was never built.

On November 10, 1885, Gotlieb Daimler and his assistant, Wilhem Maybach, first tried a motorcycle with a wooden frame and wheels, driven by a 0.5 horsepower four-stroke internal combustion engine and capable of attaining a speed of 12 kilometers per hour. From 1894 on, the first relatively consistent production of motorcycles began, thanks in particular to Hildebrandt and Wolfmüller, who manufactured a 1488 cc two-cylinder model, two thousand units of which were made. The brands multiplied in the United States and after a few years, there were about a hundred and fifty American manufacturers, the best known of which were Thor, which also built engines for Indian, Merkel, Peerless, and Yale. The immensity of the territory, combined with rapid industrial development, favored competition. In addition, copying between manufacturers led to the invention of increasingly sophisticated and technically reliable designs.

With the passage of time, however, the market stabilized, and the number of companies decreased as a result of competition, economic risks, wars, and mergers. In the end, the last fifteen great American manufacturers disappeared one after another, as a result of joint ventures, foreign takeovers, or bankruptcy due to faulty production. The consequence of this phenomenon was that only two manufacturers survived till the end of the Second World War: Harley-Davidson and Indian, which was founded in 1901, two years before Harley-Davidson. But gradually Indian was obliged to enter into partnerships and to distribute first the British brand Vincent, then others such as A.J.S., Matchless, Norton, and Royal Enfield, before it finally suspended production in the U.S. in 1953. In 1959, the Indian Company was taken over by Associated Motorcycles, the British maker of Matchless, A.J.S., and Norton. The Indian name was used through 1962. After that, several unprofitable attempts to relaunch this prestigious brand were made, most recently in 1994, when a company based in Albuquerque, New Mexico, manufactured an Indian prototype. But at the present time, Harley-Davidson alone is still functioning as an active and particularly dynamic American brand in the United States.

The United States at the Beginning of the Century

The United States has been in full economic swing since the beginning of the century. Two important factors favored industrial development: the massive influx of immigrants and the country's natural resources. Specialized and particularly well-mechanized businesses accounted for more than half of the farms; and immigrants rapidly formed an effective labor force. At that time, only one third of the population lived off agriculture, while the other two thirds were shared among industry, trade, the professions, and transport. Purchasing power as well as consumption increased so much that sale on credit developed and advertising became a very widespread promotion technique.

At the beginning of the century, the United States had almost seventy-six million inhabitants, of which twenty million lived west of the Mississippi. The 220,000-mile railroad network was longer than that of all the European countries put together.

In 1903, the U.S. automobile industry had 180 manufacturers (though this number fell to 44 within 23 years). The inevitable concentration benefitted Chrysler, General Motors and Ford, who manufactured more than 80% of all automobiles. Nearly twenty million passenger cars travelled the American roads. Air transport, too, uttered its infant's cry in 1903, with the Wright brothers' first flying tests. All these means of transportation obviously played a critical role in the economic development of the country at the beginning of the century.

The motorcycle industry experienced a concentration similar to that which occurred in the automobile industry, both in the United States and in Europe. The major difference between these two industries lay in the level of profit, which was far lower for the manufacturers of two-wheelers. But the founding fathers of the American motorcycle industry did not consider money a priority: they were idealists, passionate people who dreamed and, after visualizing their designs on a drawing board or in a sketch, built what they believed to be the ideal machine on which to cross this enormous country.

The beginning of the century seemed to justify them; in 1910, the number of motorcycles registered in the United States soared to 86,400.

16-17 center ***At the beginning of the twentieth century, trolley rails and horse-drawn carriages made it difficult to travel through a city like Milwaukee by motorcycle.***

16 bottom ***When the first motorcycles appeared, they ran side by side with the most common transportation of the time: the horse.***

17 top ***In the early twentieth century, the streets of Milwaukee, like those of any other city of the time, were used more by pedestrians than by engine-driven vehicles.***

16 top ***Louis Guillaume Perreaux-Michaux's light motorcycle was the first to be patented in the world, in 1868. It weighed 61 kg and had a small steam engine, which allowed it to achieve the speed of 15 kilometers per hour. This unique motorcycle now belongs to the Ile de France museum in Sceaux. Photo: Robert Grandseigne.***

17 bottom ***The first two-wheel engine-driven vehicles to appear on the market were bicycle frames adapted to incorporate engines.***

Friendship Generates a Legend

In 1880, in Milwaukee, William S. Harley was born into a family of British immigrants from Manchester. At the age of fifteen, he started working for a bicycle manufacturer. He had a talent for mechanics and drawing, and a few years later, at the age of twenty-one, he became an apprentice draughtsman in a metal treatment factory. It did not take long for him to show his talent for invention: being fond of fishing and wildlife photography, he conceived a system which allowed him to take shots of birds in their natural environment. In the factory where he worked, he became good friends with an old schoolmate named Arthur Davidson, who was employed as pattern-maker. Arthur's family had emigrated to the U.S. from Aberdeen, Scotland, in 1872.

Arthur and William shared the same passions: nature and everything concerning mechanics. From the automobile to the bicycle, they read all the specialized journals, which also devoted numerous articles to the new means of transport: the motorcycle. From the beginning

19 This 1910 archive photo displays the four founders, left to right: William A. Davidson, Walter C. Davidson, Arthur Davidson, and William S. Harley.

of the century, William S. Harley and Arthur Davidson had a strong desire to join the large family of motorcycle pioneers. For the time being, they did not try to come up with a name for their future creations, but rather they tried to put their ideas into practice, though they were greeted with skepticism from their peers, who believed that such a project could never succeed.

Arthur's brother Walter, while keeping his job with the railroad, also worked very actively on the launch of the company. Even though he proved to be very good at business, it is as a racing rider that he later became famous and contributed to the prestige of the brand. A third Davidson brother, William A., who joined the team as manufacturing technical manager, also played a key role.

At the beginning of their adventure these friends built three prototype engines, one of which had enough power to drive a small boat. Unfortunately, neither drawings nor elements of these three first engines have survived.

It was in about September of 1900 that the two friends undertook a study of mechanics, with a special focus on motorcycle engines. The result proved profitable; by the beginning of January, 1901, four engines were ready to be fitted on the frames of entirely classical bicycles. Their carburation was far from satisfactory, because a system for regulating the fuel supply was still to be found. Since starting is always difficult, their business remained modest both in terms of equipment and financial support, but they were determined to improve. In order to obtain the money needed for research, Arthur Davidson found an additional job and, in doing so, met another engineer, Ole Evinrude, who specialized in boat engines and would help in the development of certain parts. On another front, a German immigrant called Emile Kruger, who worked at the same factory as Arthur Davidson and William S. Harley and who had been employed by Aster Concern of Paris, had the technical drawings of the De Dion petrol engines, which were built in France. His technical knowledge, William S. Harley's experience in bicycle manufacturing and Arthur Davidson's plans to build a small air-cooled petrol engine all combined to speed up their project by rapidly solving a number of problems.

The Dream Comes True

1903: The Beginning of a Great Story

***20** The first workshop for producing Harley-Davidson motorcycles was a simple 10-by-15-foot wooden building. In 1904 its size doubled, to provide enough space to build the eight machines that the company would make that year.*

***21** The first Harley-Davidsons were painted black and decorated with gold stripes and red and gold lettering, as seen on this 1906 model. That same year, the catalog boasted a second color: the famous Harley-Davidson gray, which would be linked to the history of the line.*

Arthur's brother Walter, who was twenty-six in 1902, worked for the railroad, having started out as apprentice engine driver in Milwaukee, and then worked in Kansas and Texas before returning to Illinois. One day in Parsons, Kansas, while he was repairing a locomotive, his work was interrupted by the arrival of some mail, including a letter from Arthur inviting him to come and drive the latest prototype he'd built together with William S. Harley. Since Walter already had to go to Milwaukee for a wedding, he decided to try the first Harley-Davidson on that occasion. But the motorcycle was still in pieces when he arrived. Arthur's invitation had in fact been a lure; Arthur and William were relying on the services of a specialist like Walter to put the final touches on bike. Seduced, Walter had no difficulty accepting the invitation to collaborate on the project, but only in his free time.

The designing and manufacturing continued rapidly, despite the need to machine each part by hand, because the prototype had to be ready for tests planned for the spring of 1903. When the family house became too small for such an enterprise, a mutual friend lent them his garage and his tools (including a lathe and a press), and they were able to complete their first motorcycle, which could at last be tested. Among the various anecdotes, there is a legend saying that the carburetor of that first motorcycle was made from a tin of tomatoes, which is not contradicted by any documents, historical or current. In any case, it held the road fine, thanks to its 400 cc, 3 horsepower engine. The transmission was through a leather belt, which was compact, easy to fit, and gave a pleasingly smooth drive. Its speed, which was calculated at around 25 mph, proved quite reasonable, but the engine power was still insufficient to overcome mild inclines in one go. That "two-wheeler" also revealed to them that an engine adapted to a bicycle frame did not make a real motorcycle, because the transmission, brakes and steering were designed to cope with the strength of the cyclist's leg, not with the power of an engine.

William S. Harley sat down again at his drawing board to study what would be required for a supporting structure to replace the excessively weak frame of a bicycle. It was also necessary to change the wheel size and to

HARLEY-
DAVIDSON

22-23 Walter C. Davidson, the first president of the Harley-Davidson Motor Company, after winning the Federation of American Motorcyclists Endurance Run in 1908.

increase the bore and the engine stroke, as well as the magneto flywheel diameter, which was raised from thirteen to thirty centimeters. Every part had to be studied separately and built to withstand serious stresses, to eliminate the risk of breakage or deformation.

It was at that time that Walter Davidson, Sr. took up the dream of his brothers and William S. Harley and joined their venture. He was responsible for making the tools needed for the construction of the engine. Moreover, to give even more family character to the project, it was the Davidson brothers' aunt Janet who decorated the bike in the Harley-Davidson colors. The badge logo she created would be used until 1925. The color inside the letters would sometimes change, but the initial red would never be completely abandoned, and can still be found on some current models. When that first bike was due to be launched, its creators wondered about the final name to give to their brand: Davidson-Harley or Harley-Davidson? That the latter formula was chosen was due initially to its sound but also to the fact that William S. Harley designed that first model.

For Harley enthusiasts, the year 1903 is crucial in the story of this prestigious brand. It appeared in all the ads and was mentioned in all the articles describing this legend. In 1903, William S. Harley and the Davidson brothers introduced their first motorcycle, whose manufacture remained, by force of circumstances, confidential. Three motorcycles had been made and sold by the summer and the autumn of 1904. Remarkably, these three one-cylinder and belt-driven motorcycles were sold even before they were made.

The first of them had a career which is a famous part of the brand history: its purchaser, a Mr. Meyer, rode it for 6,000 miles before handing it to its second owner, one Lyon Georges, who rode it for over twice at distance, covering 15,000 miles. The next owner, Mr. Webster, covered the even greater distance of 18,000 miles, and that was not the end, because Louis Fluke then reached a total of 12,000 miles. Finally, the last rider of that early Harley, Stephen Sparrow, surpassed all his predecessors, by adding 32,000 miles to those already covered.

That model was equipped with a 2 horsepower, 24.7 cubic-inch engine, with a 3 x 6 inch ignition coil and a carburetor. It could run at speeds from five to forty-five mph. The transmission was still based on a 1¼ inch flat leather belt, the tank had a capacity of one and a half gallons, the wheels had a diameter of 28 inches, and braking was by back-pedalling; this model, in black, sold for $200. It was a success for Harley and the Davidsons, one which they

23 top This 1910 photo, now in the Harley-Davidson Motor Company's archives, shows the four founders, whose varying talents complemented each other.

23 bottom In 1904, only eight motorcycles were sold. In 1906, fifty machines came out of the Harley-Davidson Motor Company workshop.

did not hesitate to use as an example for advertising the qualities of their machine, the first of which, they pointed out, had covered more than 100,000 miles with its original bearings.

As for the second model, in gray, the manufacturers focused particularly on the driver's comfort and decided to eliminate the open exhaust. This bike offered such a high-quality ride that it was rapidly baptized "The Silent Gray Fellow."

After these successes, the partners were convinced that they should not rest on their laurels and that they must further develop their business. Despite material difficulties due to the lack of money, they decided to double their workshop area; they had outgrown the famous small wooden shed, which was never intended for their purpose. Thanks to these arrangements, Harley-Davidson ended up manufacturing eight "Silent Gray Fellows" in 1904. Then, when Walter Davidson, Sr. finally left his job with the railroad to join his brothers' business, it became clear that the team needed to improve its technical knowledge in order to create a more solid base for the company. This induced William S. Harley to take a course in the field of combustion engines at the University of Wisconsin, in Madison. Everything seemed poised to ensure that the small motorcycle manufacturer would not remain small for long. The choice that William S. Harley and the Davidson brothers faced at this point proved very simple: either their small factory could remain as it was, in which case it was bound to disappear one day, or they could take the risk of expanding.

The Harley-Davidson Motor Company Is Founded

We know which direction they decided to follow: on September 17, 1907, four years after the launch of its first motorcycle, the Harley-Davidson Motor Company was officially registered at the U.S. Trade and Company Register Office. It was no longer a matter of overseeing a small family workshop but of managing a real factory.

24 top Dating back to the first decade of the twentieth century, this picture shows the manufacturing team. The Harley-Davidsons they were building came equipped with a one-cylinder engine, one of which appears on the left.

24 bottom In 1907, the company enlarged its factory to meet the growing demands of motorcyclists who wished to drive Harley-Davidson models.

25 Incorporating improvements such as the addition of chrome-plated elements, this one-cylinder X8A accounted for most of the Harley-Davidson motorcycle sales in 1912. It had a 30 cubic-inch (491 cc) one-cylinder engine with a belt-tensioning lever fixed on the right of the fuel tank. The free-wheel lever, new to this model, was the ancestor of the modern clutch, its mechanism housed in the rear wheel hub. A magneto lighting system was attached to the right-hand side of the motorcycle.

In fact, while in 1904 Harley-Davidson manufactured only eight motorcycles, in 1906, fifty machines left their workshop. With such growth, it was no longer possible to remain in the small wooden shed, even though its original size had been doubled. Soon men and machines were working in a large building in Milwaukee's industrial area, along Chestnut Street (later changed to Juneau Avenue) and 27th Street. The move was largely made possible by the support of James McLay, an uncle of the Davidson brothers and a beekeeper, who agreed to finance the new building, and that of their sister, Elisabeth, who invested a substantial sum and became the owner of several parts of the business.

The new structure demanded a meticulous distribution of roles, although each person could intervene in all sectors. The management team, as it was given to the Trade Register Office at the time, was as follows:

President: Walter C. Davidson, Sr.
Vice President and Works Manager: William A. Davidson
Secretary and General Sales Manager: Arthur Davidson
Chief Engineer and Treasurer: William S. Harley

Walter and William A. Davidson were appointed to the first two posts mainly because they were the only two who had to provide for a family. In fact, as a result of the excellent relationships he was able to build with the employees, to whom his door was always open, William A. Davidson was soon considered the heart of the company. One fact illustrates the point: whenever the workers needed part of their salary in advance, he granted it to them and noted the amount in the small notebook he always had with him. But it was a pure formality; he very rarely asked for reimbursement.

His brother Walter was not happy with his post of president, and started attending a specialized training course on hot metal working.

Arthur took charge of commercial development and started to assign salesmen to several large cities: New York, Philadelphia, Chicago, Atlanta, San Francisco, and Los Angeles.

William S. Harley finished his university studies in 1908, with a degree in engineering. He then attended some courses on oxy-acetylene welding.

26-27 In 1914, Harley-Davidson offered a 35 cubic-inch (573 cc) one-cylinder motorcycle with a chain secondary transmission, fitted with a new starting system. Now the motorcycle no longer started by pedaling or by running at its side.

26 As the factory grew, machines could be assembled more efficiently, and production increased considerably. As it standardized jobs, Harley-Davidson would consolidate its reputation, diversify its production, and improve its models over the first two decades of the twentieth century.

These small details show just how determined each man was to do his best in his field, to ensure that Harley-Davidson motorcycles would not become simply ordinary motorcycles and that the brand would continue to develop.

While studying at the university, William S. Harley continued with his research to improve the one-cylinder engine that had enabled him to launch the brand and had been praised by their customers. This motorcycle, which had an initial 165 cc displacement, was improved from year to year until it was fitted with a 4 horsepower, 550 cc engine, capable of running at a speed of 36 mph.

By 1918, this model had been technically upgraded: the power of models 6A, 6B, 6C, Series 6, was increased to 30.17 cubic inches, with an increased wheel diameter, as well as a cylinder which grew from $3\frac{5}{16}$ to 4 inches and was available with one or three gears. In 1918, its manufacture was stopped, apart from a few motorcycles made on special requests, perhaps for customers looking for spare parts. After fifteen years on the scene, with 2500 units manufactured, this model, which began in a small wooden-walled workshop and had its success in a real factory on Juneau Avenue, deserved to retire.

27 top From 1909 onwards, Harley-Davidson abandoned the horse and adopted a modern truck for deliveries.

27 bottom right Diagram of the 1909 belt-tensioning system.

27 bottom left Motorcycle advertising started early in the United States, first in local and high-circulation newspapers; then, once they had been created, in specialized motorcycle journals.

November 1912 Cosmopolitan Magazine

HARLEY-DAVIDSON

The Most Comfortable Motorcycle And Why.

YOUR motorcycle must be comfortable and easy riding if it is to render satisfactory service. The fact that the motorcycle is the most rapid, practical method of transportation known, (aeroplanes and automobiles not excepted) as well as the most economical, will be more than offset if its use means positive discomfort to you. The

Harley-Davidson

is the one motorcycle that is truly comfortable. It's Ful-Floteing Seat and Free-Wheel Control (exclusive patented features) have done away absolutely with the discomforts of the ordinary motorcycle.

The Ful-Floteing Seat does away entirely with the objectionable jolting and jarring due to rough roads. It places 14 inches of heavily compressed, concealed springs between the rider and the bumps and jars. This feature alone is largely responsible for the tremendous increase in popularity of the motorcycle.

The Free-Wheel Control, a device built into the rear wheel, acts like a clutch on an auto, doing away entirely with the objectionable running alongside or hard pedaling when starting. Instead, the Harley-Davidson can be stopped and started by the mere shifting of a lever.

The comfortable qualities and other features will be willingly demonstrated by any Harley-Davidson dealer.

Descriptive literature sent on request.

HARLEY-DAVIDSON MOTOR CO.
257 B STREET MILWAUKEE, WIS.

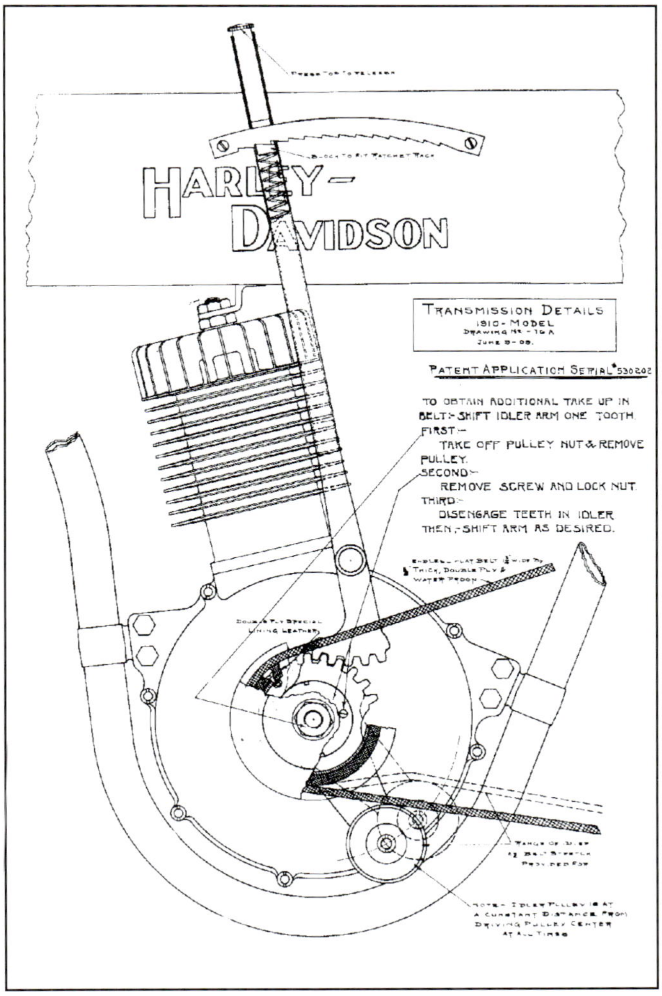

HARLEY-
AVIDSON
HARLEY-DAVIDSON
MILWAUKEE-U.S.A.

The V-Twin Era

The Engine That Built the Legend

The first Harley-Davidson V-Twin appeared in 1909. One important detail must be mentioned here: this engine was not invented by the American company, and Harley-Davidson was not alone in adopting it. But the company kept and improved it, and it contributed to the brand's image over time. Thanks to its shape – an upside-down triangle – this engine fitted the bicycle frame, as was the case of the Harley-Davidson Motor Company "5D" model, in 1909.

Engineers easily established that an engine built with only one cylinder was naturally limited, mainly in its displacement and therefore in power, in such a way that it could withstand neither stresses nor an excessive efficiency.

The single cylinder, in existence since 1903, could be inclined both forwards and backwards. This positioning left room for a second cylinder, which in turn increased the possibilities of the engine, by keeping a single reinforced crankcase and fitting a modified timing distribution system control shared by both the cylinders.

Harley-Davidson adopted a forked connecting rod for this first model, which was fitted with the famous "V-Twin," thus doubling the engine displacement and obtaining greater efficiency. The two cylinders were set at 45 degrees, with an inlet valve on the head and an exhaust valve at the side. This two-cylinder engine had a displacement of 850 cc (50 cubic inches), whereas the one-cylinder model did not exceed 500 cc. This new model had all cable controls: the accelerator and the spark advance were located at the level of the grips. The 1909 "5D" model had a magneto and a 28-inch wheel, which could be requested as an option.

The "5D" model reached 62 mph with neither clutch nor gear box. In comparison with the one-cylinder model, the two-cylinder model kept pedals and a bicycle chain on the right, as well as a leather driving belt on the left, but its horsepower was increased by three.

Yet there were several technical handicaps: there was no tensioning pulley, the leather belt was inclined to slip, and the rider had to turn

28 The first prototype of the Harley-Davidson V-Twin was unveiled during a motorcycle show in 1907, but the model was not actually produced until 1909. Design flaws slowed production, but from 1911 on, the V-Twin reappeared in the catalog and rapidly met with success.

29 top Head of the engineering department, William Harley proudly presents the first page of the **Dodge City Daily Globe*****, announcing Harley-Davidson's victory in a 200-mile race.***

29 center Since its first configuration in 1909, the Harley-Davidson V-Twin adopted a new exhaust outlet called the "pocket valve," which replaced the atmospheric valve taken from the old De Dion engines. V-Twin production stopped in 1910, but later a few modifications made these valves reliable and efficient. Shown here is a personalized machine from the mid-1920s.

29 bottom As early as 1914, the Harley-Davidson Motor Company promoted its motorcycles through advertising, boasting the merits of its models and attracting more customers by establishing specialized dealers who were able to maintain its machines.

off the engine every time he wished to stop his motorcycle.

This did not prevent the "5D" model from launching Harley-Davidson on the path the company has followed for almost 100 years.

In 1909, when the newly graduated engineer, William S. Harley, finally returned to the factory, he was convinced that production could only benefit from his scientific and technical knowledge. It was actually in 1908, on his drawing board, that he had conceived, in theory, the first two-cylinder engine. At first glance, it seemed that one had only to add a cylinder to the existing engine, but experiments showed that the new engine had to be particularly solid to withstand vibrations and other damage caused by the poor state of the roads, especially over the long distances which motorcycles might cover. As mentioned earlier, this engine, which went into production in 1909, had two cylinders positioned at 45 degrees and valves placed at the sides; it also generated greater torque than the one-cylinder model, especially at low speed. As with the one-cylinder model, the engine was extremely robust thanks to heads which were an integral part of the cylinders, and an inlet valve with an external rocker arm, which was articulated on the cylinder head. This allowed the rotation speed and engine compression to be increased. Even in its first tests, the new engine generated very large vibrations; this phenomenon would become part of the Harley-Davidson legend.

In 1910, the Harley-Davidson production catalog included, in Series 6, three 30.17 cubic-inch one-cylinder engines, the 6D V-Twin with 49.48 cubic inches, 6.5 horsepower, and 28-inch wheels. In 1911, Series 7 consisted of the same models (references: 7A, 7B, 7C for the one-cylinder and 7D for the V-Twin). In contrast, Series 8 of the following year offered four one-cylinder and three two-cylinder models, the 8D, the X8D, a 49.48 inch displacement, and the X8E with 60.61 cubic inches and with 7 to 8 horsepower. This last model, which was supposed to be a sports model and which could only be driven by expert drivers, was the only one to have a clutch and a free rear wheel. On this V-Twin, entirely researched and designed by William S. Harley, a chain, which was tensioned thanks to a lever located on the left side of the tank, replaced the original leather belt.

The year 1912 marked the release of the geared V-Twin engine, where a chain-tightening lever was modified for this use and could also operate as a gear lever.

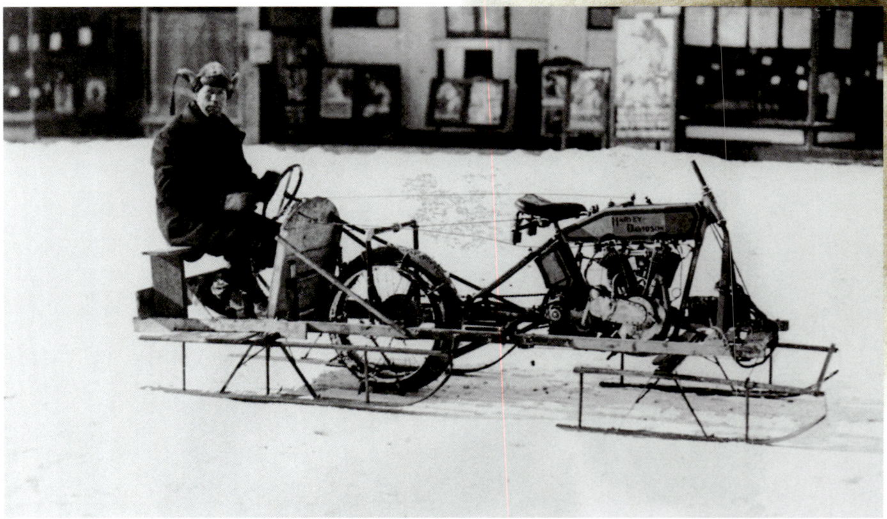

30 top left The "Motorcycle Truck," later called "triporteur," reached the market in 1915. It met with huge success and was used by a wide variety of customers.

30 top right Keen on its brand image, Harley-Davidson developed a magazine for dealers and committed itself to advertising through quality catalogs.

30 bottom The imagination of Harley-Davidson enthusiasts proved ebullient, which even surprised company officials. Take for example this 1917 ancestor of the skidoo.

30-31 and 31 top The "Motorcycle Truck" carried a large wooden trunk with a top cover, fixed on a directional chassis. It was driven with a handlebar, like a motorcycle. These machines were praised as a practical way to make deliveries. Less expensive than cars or trucks, they ended up providing excellent company publicity. Harley-Davidson manufactured them on order, with paintwork which included the logos of the companies purchasing them.

32-33 This is how motorcycles sent by the Harley-Davidson Motor Company were packed in 1916.

That year, another great novelty was introduced to customers: the seat was no longer a simple bike saddle with its two vertical spiral springs, but a real seat with a shock absorber fitted inside a tube which was an integral part of the frame.

In 1913, Harley-Davidson created its own racing service. Technical responsibility was entrusted to William Ottaway, who previously worked in the motorcycle department of Aurora Automatic Machine Company, in Illinois. Ottaway was also a racer, and when he went to work for Harley-Davidson, he had a reputation as a winner and an innovator in the motorcycle field. It was hoped that Ottaway's qualifications and his racing experience, together with William S. Harley's knowledge, would benefit the Milwaukee factory. He worked on technical improvements, specifically, the implementation of new systems which would allow the engines to perform better and to further withstand stress. In particular, his research resulted in an increase in the engine speed. He also conceived a new gear-changing mechanism, which then enabled Harley-Davidson motorcycles to move heavier loads, such as passengers in a sidecar. These early modifications appeared on the 1914 Harley-Davidson models, such as the "10B," which was equipped with a single cylinder and a single gear, and the "10C," which is the only one-cylinder engine that has two gears. In the same year, a V-Twin engine, referred to as "J," with 61 cubic-inch displacement, had two gears and a chain transmission. Ottaway permanently improved the qualities of this engine, increasing its power by 30%. In 1915, he implemented a foot-start system by means of a kick lever, eliminating the need to pedal.

During this period, the motorcycle became increasingly popular, thanks especially to the sidecar, which became widely used. Sidecars were produced by the Rogers Company, an American manufacturer which sold 2,500 of them to Harley-Davidson in 1914 and which doubled its production over the following year. Americans were discovering a new, reasonably priced means of transportation for more than one person. In 1915, these motorcycles were fitted with components still used today, such as a choke for cold starts, two-gear transmission, and back brakes.

In that period, single-cylinder engines enjoyed their last hours of glory. Over the years, the technology had improved to 30.17 cubic-inch displacements and 4.34 horsepower, but it was time for this model to be consigned to memory and to make room for new models which would forge the future of the Milwaukee motorcycle. During this time, the Harley-Davidson Motor Company continued its efforts to improve the revolutionary 1909 V-Twin engine. Bill Ottaway had given it a displacement of 61 cubic inches, or 1000 cc, which was very impressive for that time. The technological developments achieved in racing would continue to benefit production models. In 1917, Harley-Davidson put Model "17" on the market. A real bench mark for Bill Ottaway, this bike had an engine with four valves per cylinder and unique 54-inch wheels. Unfortunately, it was far from perfect: it was not fast and the rider had to lean either forwards or backwards, according to his physical build, in order to balance it. However, by increasing the caster angle, Bill Ottaway found a solution to this problem.

Two years later, a 37 cubic-inch engine, the "Sport Model," was launched on the American market. Fitted with a four-stroke engine, it had a displacement of exactly 584.02 cc (6 horsepower) and could rotate at up to 4500 rpm, thanks to ball bearings on all the big ends. The engine flywheel was closed inside a sealed case. This three-gear Flat-Twin motorcycle, which had a chain transmission that was set by removing the back wheel, was manufactured between 1919 and 1922. In August of 1919, during a 1,012-mile run from New York to Chicago, it distinguished itself, ridden by Julian C. "Hap" Scherer, who covered the distance in thirty-one hours, twenty-four minutes.

33 top In 1918, Harley-Davidson became the most important motorcycle manufacturer in the world. Its success was due particularly to the V-Twin, which proved to be suited to being riden alone or with a sidecar, which was then quite popular.

At that time, most of the motorcycles manufactured in Milwaukee were exported: war and the economic situation on the continent had forced European countries to reduce or even stop their national production. This was the case of England's Triumph, a company which had been founded in 1903, too, by two Germans, Siegfried Bettmann and Maurice Schulte. Since 1912, though, their 548 cc single-cylinder engines were manufactured by a branch based in Detroit.

Harley-Davidson continued at the same pace and won further endurance races, with respectable records for each of them. Credit for the most spectacular of the trials must be given to Edwing Hogg, who crossed Death Valley. For his race across this 150-mile desert, Edwing Hogg chose a Sport-Twin, which was thoroughly overhauled by the Harley-Davidson service department before his departure. He crossed Death Valley a number of times, during which his only complaint was the large number of punctures. Once more, the Sport-Twin and Harley-Davidson had lived up to their reputations.

In 1920, motorcycle manufacture increased by almost 4,000 and reached the sum of 27,040 machines. Harley-Davidson motorcycles had been fitted with a magneto dynamo that allowed the battery to recharge when running, which was impressive, but the most impressive improvement of this era was the fact that the motorcycle could start even if the battery was dead.

33 bottom William A. Davidson and William S. Harley, on their way back from a fishing party (a sport they both particularly enjoyed), realized their combination of a Harley-Davidson and a sidecar was an ideal vehicle for leisure rides and short journeys.

The Company's Technical and Commercial Policy

The success of the V-Twin and the growth of production spurred the managers to increase the number of sales people. In 1910, this led to the creation of a proper network throughout the United States.

Arthur Davidson was personally in charge of selecting the future dealers according to very strict criteria and making in-depth investigations to determine their professional qualifications, reputation, and inclination to build up good customer relationships. Since most of them were bicycle dealers scarcely competent in mechanics, Harley-Davidson provided them with training in the new field so that they could better inform their customers and ensure good after-sales service. Some of the early salespeople, such as C. H. Lang, belong to the Harley-Davidson legend. Lang was one of the brand's first dealers. From 1912, he alone was selling 800 motorcycles

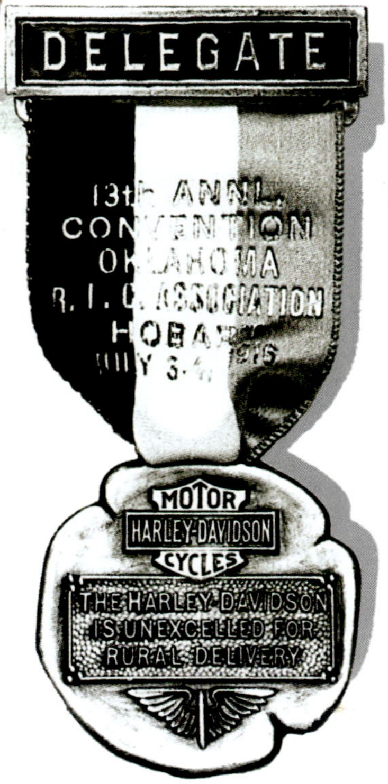

***34 top** A good-looking girl, a Harley-Davidson, and a romantic drive – what else could one desire to be happy? This Harley-Davidson cliché still exists today.*

***34 bottom** Harley-Davidson received a medal for creating vehicles that promoted communication between rural areas.*

***35 top** Harley-Davidson dealers became more and more important. The character of their operations contributed to the company's image, especially in terms of service long after the motorcycles had been delivered.*

per year. (Total production in 1912 was 3,852 motorcycles.) In order to attract customers, he introduced a very appealing credit system. Thanks to his aptitude for trade, his activity did not slow down, and he became, in the space of four years, the most important Harley-Davidson dealer in the whole country.

But Arthur Davidson was not content with a network of dealers only inside the United States, so he built other networks in Australia, New Zealand and Europe (in England, Italy, France, Germany, Spain, Sweden, Denmark. . .). By 1921, Harley-Davidsons could be purchased in sixty-seven countries around the world. Some European manufacturers, seeing the public's interest in the motorcycles from Milwaukee, did not hesitate to copy the brand. After the Great War, the French Harley-Davidson importer

***35 bottom** Harley-Davidson has been publishing a magazine,* **The Enthusiast**, *to promote the company since 1916.*

***36 top** Sidecars enjoyed considerable tourist and commercial success, due mainly to their price as a family vehicle, as compared to the price of a car. Furthermore, trunks from commercial models were directly attached to the family sidecar chassis. By replacing them with baskets, a family could easily pack up for a weekend ride.*

launched a small 123 cc single-cylinder motorcycle which looked like the "Silent Gray Fellow." Produced from 1923 to 1928, it was sold in France under the name of "Harlette-Geco."

To improve its commercial performance and set itself apart from its competitors, the Harley-Davidson Motor Company began publishing several magazines. In 1912, the brand created *The Dealer*, a report to bind the company and its dealers, which was published until 1916. That same year, the company launched *The Enthusiast,* available in English and Spanish, which kept dealers and customers informed of all the novelties coming from the factory. This paper also dedicated a few pages to miscellaneous reports which were in some way connected to motorcycles; it is still published today.

Harley-Davidson, like other manufacturers, also advertised in the first U.S. magazines aimed at fans of this increasingly popular sport: *Bicycling World, Motorcycle Illustrated* (published in Illinois), *Motorcyclist* (Los Angeles), and *The Western Bicyclist*. All of them reported on the sporting events taking place around the country and presented the new models and accessories. It is worth pointing out that the motorcycle was an increasingly popular means of transportation in the United States until 1908, when Henry Ford launched his famous Model T.
Since car production was still very limited, automobiles were a lot less affordable than motorcycles: a motorcycle sold at $250 to $290, whereas it took $900 to buy a Ford T. The drawback of the motorcycle was, obviously, the smaller number of seats, but it was possible to fit a sidecar for less than $100, which explains the success enjoyed by these vehicles.

In 1920, after seventeen years of existence, thanks to their technical and commercial performance, Harley-Davidson outstripped Indian, which had been America's premier manufacturer until then. An investment of $3.5 million allowed the company to further expand the factory on Juneau Avenue (to 542,258 square feet) and to equip it with modern machinery. Thanks to the improvements, the factory, with 2,400 employees, was able to produce more than 30,000 motorcycles per year.

***36 bottom** During the 1930s, Harley-Davidson-based commercial vehicles diversified, in particular thanks to the Servi-Car.*

***37 top** The Servi-Car became enormously successful with the American law enforcement authorities. The police used them for their urban corps: in the Servi-car, two policemen could ride together, one marking wheels to check parking duration of cars along the street.*

***37 bottom** The Harley-Davidson sidecars won many races, not only in the United States but also in various other countries. Their legendary reliability encouraged many teams to undertake dangerous and spectacular journeys. Ordinary people, too, used them daily to get through any weather and across any type of ground.*

38 The Dudley Perkins dealership opened its doors in San Francisco in 1914. Today, it represents the oldest Harley-Davidson concession in the world. Still managed by the Perkins family, it has become legendary. It has changed as the brand has changed, responding to the market and to riders' demands, supplying motorcycles and all the accessories, clothes, and gadgets one can imagine, all marked with the Harley-Davidson logo.

39 New Harley-Davidson dealerships, as well as those who have been operating for years, have to stay sensitive to the changing brand image. As a result, although the exterior of each shop is personalized, inside the customer finds himself in familiar territory. The layout of each dealership is familiar, but business style can be dictated by the personality of the dealer.

40 top To support sales, the Harley-Davidson Motor Company has published several magazines. In 1912, it launched **The Dealer**, *a bulletin intended for concessionaires, sold until 1916. In 1916, it introduced* **The Enthusiast**, *which continues to keep dealers and customers informed. The company has consistently used advertising to communicate with future customers, placing ads not only in newspapers but also in new specialty magazines.*

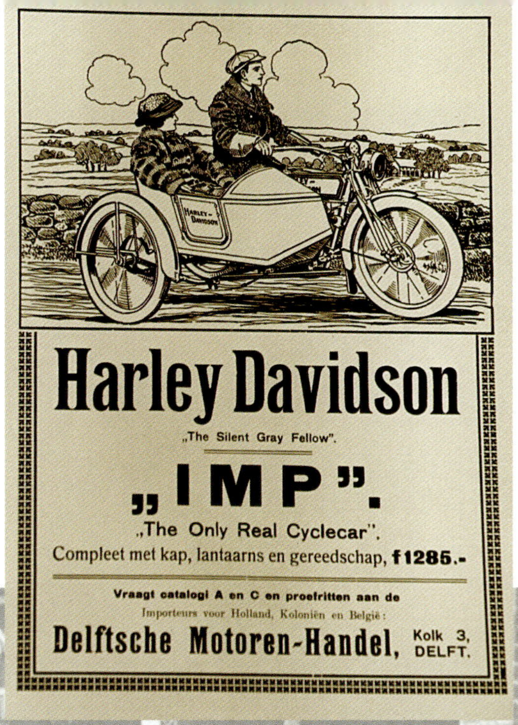
Harley Davidson
„The Silent Gray Fellow".
„IMP".
„The Only Real Cyclecar".
Compleet met kap, lantaarns en gereedschap, f1285.-
Vraagt catalogi A en C en proefritten aan de
Importeurs voor Holland, Koloniën en België:
Delftsche Motoren-Handel, Kolk 3, DELFT.

40 bottom Harley-Davidsons, like all the other motorcycles of that time, developed little by little. From simple utility machines used for transport, they gradually became leisure vehicles appealing to younger people. Racing played an important role in this process.

40-41 Harley-Davidson executives, as well as their heirs, had to be motorcycle enthusiasts. While they were all still young, Gordon, Walter, and Allan Davidson made a stop at the Dudley Perkins dealership in San Francisco in 1929.

***41 left** Harley-Davidson promotions boast about the qualities of each machine and give information about the development of the company.*

***41 right** Motorcycles and planes often appeal to the same enthusiasts. Here, Harley-Davidson focused on this theme, comparing the sensations of the pilot with the driver. Many plane pilots of the time actually traveled by motorcycle when they were on land.*

42 top Whether made by Harley-Davidson or another company, sidecars became very successful during the first part of the 20th century, not only because of price and convenience in carrying a passenger, but also because they made it safer to drive in poor conditions or when it rained or snowed.

42 bottom *The Flat-Twin Sport model, introduced in mid-1919 as Twin Sport, had a 35.6 cubic-inch capacity. From 1920 on, Harleys were fitted with an electric system and, in 1921 (the model shown here), with a new fuel tank whose design resembled that of the Big Twins. This model, which became the Electric Sport Model, was only manufactured until 1922, since it did not give Harley-Davidson a competitive edge against the sportiest Indian machines.*

42-43 *In 1920, the Harley-Davidson Company was the biggest motorcycle producer in the world, measured both in size and in the number of motorcycles produced per year. In 1926, the Harley-Davidson factory on Juneau Avenue had been a monument of Milwaukee for a decade.*

43 bottom *Exploded view of a Harley-Davidson 74 cubic-inch V-Twin (1200 cc), produced in 1924. The first 74 c.i. V-Twin engine was unveiled in 1921 and appeared on the market in 1922. At the same time, a 61 cubic-inch (1000 cc) model was in the catalog as well.*

44 top and 44 bottom ***Fitted with baskets adapted to specific jobs, Harley-Davidsons with sidecar frames proved great commercial successes, performing jobs that ranged from simple deliveries to servicing broken-down vehicles.***

44 center left ***In their marketing message, Harley-Davidson emphasized that their motorcycles were the perfect answer to customer needs of the time. It was possible to ride solo, or a leisure or commercial sidecar could easily be attached, simply by replacing the boot on the frame.***

44 center right ***This fork head, appearing in the Harley-Davidson catalog during the first half of the 1920s, became popular as a screen against bad weather. While it could affect road handling when riding solo, it proved excellent when the motorcycle was coupled to a sidecar.***

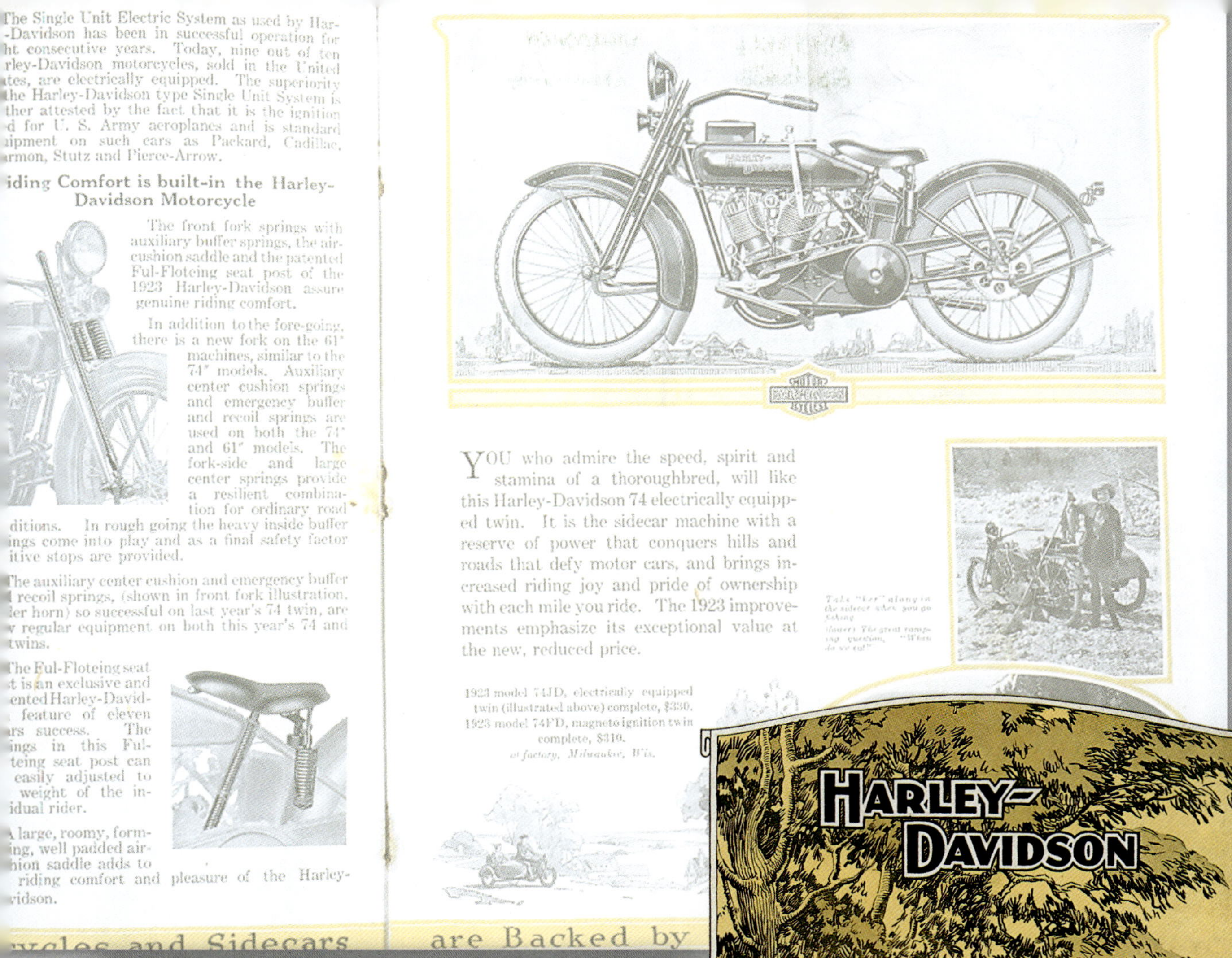
he Single Unit Electric System as used by Har-
-Davidson has been in successful operation for
ht consecutive years. Today, nine out of ten
rley-Davidson motorcycles, sold in the United
ites, are electrically equipped. The superiority
the Harley-Davidson type Single Unit System is
ther attested by the fact that it is the ignition
d for U. S. Army aeroplanes and is standard
ipment on such cars as Packard, Cadillac,
rmon, Stutz and Pierce-Arrow.

iding Comfort is built-in the Harley-Davidson Motorcycle

The front fork springs with auxiliary buffer springs, the air-cushion saddle and the patented Ful-Floteing seat post of the 1923 Harley-Davidson assure genuine riding comfort.

In addition to the fore-going, there is a new fork on the 61" machines, similar to the 74" models. Auxiliary center cushion springs and emergency buffer and recoil springs are used on both the 74" and 61" models. The fork-side and large center springs provide a resilient combination for ordinary road ditions. In rough going the heavy inside buffer ings come into play and as a final safety factor itive stops are provided.

he auxiliary center cushion and emergency buffer recoil springs, (shown in front fork illustration, ler horn) so successful on last year's 74 twin, are regular equipment on both this year's 74 and twins.

he Ful-Floteing seat t is an exclusive and ented Harley-David- feature of eleven rs success. The ings in this Ful-teing seat post can easily adjusted to weight of the in-idual rider.

A large, roomy, form-ing, well padded air-hion saddle adds to riding comfort and pleasure of the Harley-vidson.

YOU who admire the speed, spirit and stamina of a thoroughbred, will like this Harley-Davidson 74 electrically equipped twin. It is the sidecar machine with a reserve of power that conquers hills and roads that defy motor cars, and brings increased riding joy and pride of ownership with each mile you ride. The 1923 improvements emphasize its exceptional value at the new, reduced price.

1923 model 74JD, electrically equipped twin (illustrated above) complete, $330.
1923 model 74FD, magneto ignition twin complete, $310.
at factory, Milwaukee, Wis.

ycles and Sidecars are Backed by

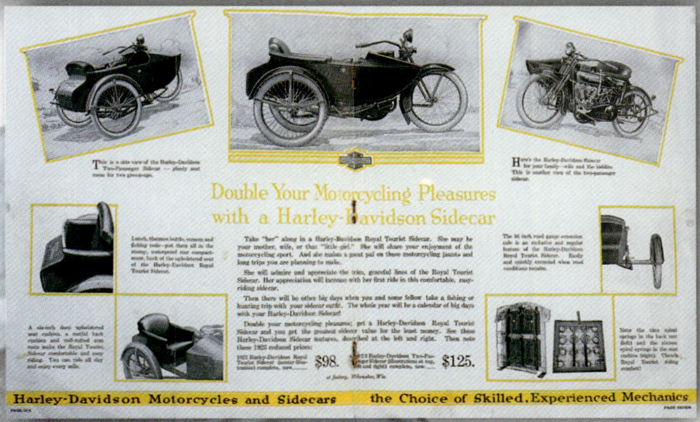

45 In 1922, the first Harley-Davidson 74 cubic-inch (1200 cc) V-Twin was launched. This motorcycle's mission was to boost sales by offering greater power than the 61 cubic-inch model. While the 61 was better for riding solo, advertising promised that the 74 could generate 18 horsepower, making it the ideal engine for driving sidecars and carrying a passenger.
A strong market demand prompted manufacturers to develop sidecar machines and improve them over the years. Sidecars had to compete with cars produced in assembly lines. In 1923, Henry Ford launched a basic version Model T at about $300, with a devastating impact on the motorcycle industry. In 1922, a Harley-Davidson JDS Sidecar Twin sold for $390. In 1923, the Harley-Davidson JD Solo was offered in the catalog at $330.

46 top Whether solo or with a sidecar, the Harley-Davidson was supposed to convey the spirit of discovery. The company wanted to convince the public that it was the ideal machine for traveling reliably, whether to the far corners of the earth or around the block, which is where most owners would drive it.

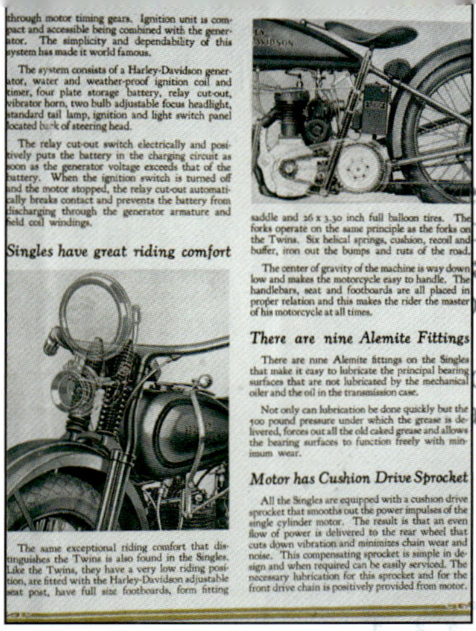

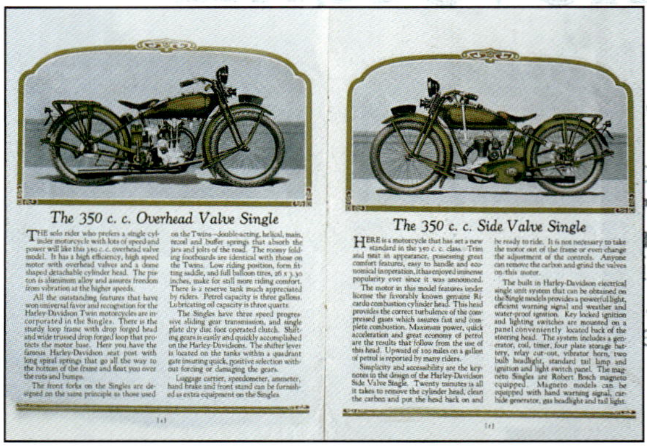

46 center and bottom A communication medium was needed – to convey the Harley-Davidson image, to portray the traveling spirit linked to these motorcycles, to inform potential owners of sports model performances, and to spread the news about innovations and improvements. Of course, the company already communicated with the public through specialized journals and high-circulation newspapers, but it needed its own publication. ***The Enthusiast*** *appeared for the first time in 1916.*

46-47 ***The Enthusiast*** *addressed both dealers and users. It began by broadcasting Harley-Davidson's military contribution during World War I. Still published quarterly today, it is distributed by dealers.*

HERE'S th... to 61 twin at a ... sets a ... tor-

1923 model 61F, magneto ignition twin (illustrated above) complete, $285, *at factory, Milwaukee, Wis.*

Motorcycling through Balboa Park, San Diego, Cal.
(lower) Lawbreakers have no chance with this West Hoboken, N. J., well-manned trio of motorcycles and sidecars.

than 900 Police Departments

THIS 1923 Harley-Davidson 61 electrically equipped twin is a greater value at a greatly reduced price. The 61 electric model is now equipped with the double-action, shock-absorbing front fork springs, silchrome exhaust valves and all the other big 1923 new improvements. In addition to its ability to pull a sidecar, this 61 model is winning more friends, each day, as a solo mount, through its ease of handling.

Go to that fishing paradise with a sidecar outfit.
(lower) Victor and victim homeward bound.

1923 model 61J, electrically equipped twin (illustrated above) complete, $305. *at factory Milwaukee, Wis.*

47 Before the Servi-Car, police officers used the Harley-Davidson sidecar, since it allowed the rider to be ready to take action, to open fire if necessary, or simply to mark the tires of parked cars.

***48 top** Finding a one-cylinder Harley-Davidson in this condition is quite rare. The collector has chosen to preserve this one in its original state rather than renovating and updating it.*

***48 bottom left** The first sidecars were used like cars, and they adopted a similar look. Early marketing messages tried to ensure the comfort of a sidecar, since they targeted people who wanted a car but could not afford one.*

***48 bottom right** Certain collectors – along with those who simply own and love their old Harley-Davidsons – still ride their classic bikes for pleasure. They also take part in rallies and shows. Sometimes, as here, they make real journeys.*

***49** It is worth noting the valve systems and the position of the plugs, features which were applied to the other American motorcycles of that time. The famous scoops in the tank permitting the passage of the inlet valve were later used by certain modern designers for aesthetic reasons when they transformed their Harley-Davidsons.*

HARLEY-
DAVIDSON

Harley
Davidso
MILWAUKEE U.S.of A.
1918
Ladies
Standard

Harley-Davidson Bicycles

The war created a demand for bicycles, too, and spurred the Harley-Davidson Company to produce them starting in 1917. Since this kind of two-wheeler was no longer imported from Europe, which was at war and whose industry suffered as a result, a complete line of Harley-Davidson bicycles was put on the market from 1917 to 1921 including, apart from the usual models for men and for racing, models for women, young girls, and children. There was also a special version for boys of ten to sixteen years of age, as well as a bicycle with a reinforced frame to which it was possible to add an engine – in other words, a return to the origin of the motorcycle. These bicycles, manufactured entirely by hand, were offered at prices ranging from $30 to $40; however, after the war, the Harley-Davidson Company focused exclusively once again on its main activity: the motorbike. Once the war was over, the demand for Harley-Davidson bicycles decreased, and the company decided to abandon this area, concentrating their investment instead in the development of their racing models and those for the average consumer.

50-51 In 1919, in order to attract young customers, Harley-Davidson started manufacturing bicycles. The company made the most of its brand image, banking on the reliability, quality, and patriotic image of their motorcycles to sell bicycles. Similar to their "big sisters," the motorcycles, Harley-Davidson bicycles were painted army drab. Their quality was excellent, but production stopped in only a matter of years because the retail price was too high.

Harley-Davidson Puts on Its First Fighting Uniform

Having begun to build up his sales network, Arthur Davidson turned his attention to the various public service departments whose employees might be likely to use his two-wheelers for their work. In 1909, during an annual postal conference, he presented a motorcycle designed to speed up the distribution of mail. A contract was agreed to and signed, as was one with the telephone company, which ordered 700 motorcycles, not to mention the police, who would use this new means of transportation to maintain law and order.

In 1916, Harley-Davidson launched its first army motorcycles in a conflict on non-European soil, during a punitive expedition led by General John J. "Black Jack" Pershing against the Mexican rebels led by Pancho Villa. Since Mexico's President Francesco I. Madera lacked the military force to face the rebels, he authorized the Americans to enter Mexico to put an end to Pancho Villa's uprising. General Pershing's troops failed, but this was not a defeat for the Harley-Davidsons, which had been tested for the first time on a real battlefield and had performed successfully.

In 1914, war broke out in Europe. Initially, the United States kept its distance from the conflict. But it was difficult for President Woodrow Wilson to remain neutral for long. The Germans torpedoed American ships during submarine fights, ignoring any right to neutral navigation. The United States declared war on Germany on April 6, 1917, and sent in two million soldiers under General Pershing. The United States also granted $10 billion of credit to the Allies. During these difficult years, Harley-Davidson equipped a number of its machines with

military equipment. In particular, William S. Harley was put in charge of studying the military's needs and adopting the necessary armaments for the missions the two-wheelers were to undertake on the European field. Twenty thousand motorcycles were used in this conflict. Although not all of them came from Harley-Davidson, the participation of the brand was far from negligible.

During this period, the Harley-Davidson Motor Company established the Harley-Davidson Service School (which closed in 1941) to train mechanics to repair, in situ, the breakdowns that were the result of hard and demanding use. Beginning in 1917, the school soon stopped accepting Harley-Davidson dealers in order to reserve places for war recruits. Almost 300 soldiers attended the school before the armistice on November 11, 1918.

These military motorcycles, which became renowned in Europe in only a few years, allowed Harley-Davidson to gain an excellent reputation. Seven thousand of them were sent to France, and certain collectors still have them today. The models reserved for France had a simple ignition magneto and acetylene lighting supplied by a compressed gas bottle placed in the center of the handlebar.

Before the war, Harley-Davidson had used German electrical systems made by Bosch, still a well-known brand today. But once the conflict started, it was no longer possible to equip motorcycles with this system; an American manufacturer was put in charge of devising an equally good system. In this way, engineers were forced to manufacture technologically excellent parts, which would benefit the whole automotive industry.

52-53 Public authorities, like private companies, rapidly replaced horses with motor vehicles, among which sidecar motorcycles gained a good market share. The U.S. Post Office even published a stamp bearing the image of a Harley-Davidson. Police, who used Harley-Davidsons to preserve the order, posed for pictures like this one.

54 top In 1917, one of Harley-Davidson's military contributions was to train soldiers to be mechanics, specializing in on-the-spot motorcycle repairs.

A CORPORAL AND HIS MECHANICAL MOUNT

On November 8, 1918, a few days before the armistice, while the German soldiers were marching towards their borders, an American messenger, Corporal Roy Holtz, driving a motorcycle with a sidecar, led his captain on a mission to the north of Belgium. The bad weather and the fighting forced Holtz to travel by night, and, having lost his sense of direction, he headed east. Attracted by the lights of an isolated farm, he stopped to ask for directions, only to discover that the farm was in fact a German headquarters. The two Americans were taken prisoner. Luckily, their imprisonment lasted just two days, for the armistice was declared on November 11. Once free, the captain and Corporal Holtz split up. The corporal headed west on his bike and met the German troops who were retreating. Thus, Roy Holtz was the first American soldier to penetrate into Germany. . . seventy-two hours before his American friends.

On November 12, 1918, somebody living on the banks of the Rhine took a photograph of a man on a motorcycle, but establishing the motorcyclist's identity was very difficult. It was a full twenty-five years before Roy Holtz (who in civilian life was an electrician near Milwaukee) was identified as the man in the picture. In 1943, the photograph appeared in *The Enthusiast,* and in 1944, Holtz went to the Harley-Davidson offices to get a copy of the picture, which had been published in many American newspapers and had become a famous image, symbolic of the end of the war.

In 1917 and 1918, Harley-Davidson produced 45,230 models. The factory manufactured five different commercial models, two single-cylinder and three V-Twin, of which one type had one gear and the other two types, F18 and J18, had three gears and a displacement of 60.34 cubic inches.

Unfortunately, the war brought hard times to the American motorcycle industry and some small manufacturers, unable to obtain certain parts from Europe, were forced to shut down.

54-55 On the 12th of November, 1918, after an amazing but true experience which he shared with his captain a few days before, Corporal Roy Holtz was the first American to penetrate Germany after the end of World War I.

54 bottom Whether ridden solo or with a sidecar, Harley-Davidsons offered soldiers significant advantages over horses.

55 William S. Harley, together with soldiers, tested the performance of motorcycles built for World War I.

From 1904, Harley-Davidson motorcycles took part in many races, but unlike other brands such as Indian, Pope, and Thor, the factory contributed neither to the preparation of race models, nor to their financing. Yet Harley-Davidson benefited from the commercial implications of the numerous victories won by their bikes. In 1913, for the company's tenth anniversary, William S. Harley finally decided to set up a racing service.

President Walter himself rode a Harley-Davidson to victory in an endurance race lasting two days. It was the first official American race organized by the Federation of American Motorcyclists (F.A.M.), which would later become the American Motorcycle Association (A.M.A.). In 1908 there was a two-day endurance race that began in the Catskill Mountains and ended in Brooklyn. There were eighty-four participants and twenty-two brands represented, but half the drivers and their motorcycles were eliminated during the first

***56 top** Harley-Davidsons rapidly managed to dominate dirt-track races, but the company didn't establish a racing department until 1914. Until then, these races were won by avid riders who had modified their machines by themselves.*

***56 bottom** Ralph Hepburn, astride an 8-valve Harley-Davidson after a series of victories. Hepburn won the 100-mile race in 1:07:05.4, the 200-mile race in 2:07:54, and the 300-mile race in 3:30:03, all at Dodge City, Kansas, on July 4th, 1921.*

day of the race. Walter Davidson managed to stay the course without difficulty and won the race with an entirely mass-production single-cylinder motorcycle.

After this first endurance event, Walter Davidson took part in another race, the Economy Run, which began on Long Island, New York. He won again. His victories boosted sales, and the publicity surrounding him resulted in an increase in the number of dealers. On Independence Day, July 1914, Harley-Davidson's motorcycles won 23 races. Leslie "Red" Parkhurst won the One Hour National Championship in Birmingham, Alabama. In 1915, a rider named Floyd Clymer won the World Dirt Track Title on a two-cylinder model fitted with eight valves and two exhaust outlets per cylinder prepared especially for the race.

While the war was raging in Europe, Harley-Davidson and Bill Ottaway continued to win victories on the race track. They also organized a race in Venice, California, at which the main motorcycle brands would compete. The best drivers were there, and Bill Ottaway, a master of organization, chose drivers capable of taking the top positions.

***56-57** Board track races became very popular in the 1920s. Many board track riders became famous, and many spectacular accidents took place on these oval wooden tracks. Builder Jack Prince specialized in the construction of these wooden tracks from 1908 to 1925.*

***57 top** Eddie Brink in 1927, astride the "Peashooter," a 21 cubic-inch (350 cc) one-cylinder Harley-Davidson. In the same year, Brink died on a board track, by falling in front of Joe Petrali.*

58 After World War I, Harley-Davidson started competing officially. Despite some early successes, the racing department had to work hard to match its most threatening competitors, Indian and Excelsior, pooling the efforts of many technicians in support of the riders.

58-59 In the early part of the century, renowned riders took part in motorcycle races as members of teams. They went on tour throughout America. The most successful of them drew wages from the manufacturers. In general, they started up their own careers, then gained the support of local dealers before the national companies themselves agreed to sponsor them.

Among the great drivers hired by Harley-Davidson were Otto Walker, who won the race with an average speed of 68.31 mph, and Red Parkhurst, who was ranked second. In taking on Otto Walker, Harley-Davidson made the right choice, for Walter consistently won races, with an average speed ranging from 79.84 to 89.11 mph. In 1915, Harley-Davidson motorcycles won twenty-six victories.

In 1916, Harley-Davidson built its first circuit monster for a group of drivers nicknamed "The Wrecking Crew," namely Hepburn, Cunningham, and Parkhurst. In that year, the motorcycle, which, incidentally, cost $1,500, won many races, including fifteen national championships. This first sport model, fitted with eight valves and boasting a displacement of 61 cubic inches, three gears, and a very long frame, was available in three versions. Stripped of the front mudguard, it was fitted with a spring solo-seat located very low, which lowered the center of gravity, and its V-Twin engine was positioned centrally. Apart from being short and curved towards the ground, the handlebar was parallel to the front of the frame. The wheels measured 51 inches. It consumed nearly a gallon of oil, around three gallons of fuel, and was unstable at speeds greater than 60 mph.

After a number of tests, this racing bike was redesigned and Harley-Davidson once again faced the other great American brands. In Dodge City, Kansas, thirty-six drivers left the starting line in a extraordinary heat, but as a result of the 11K's mechanical problems, Harley-Davidson forfeited victory in this race which remains a classic among motorcycling events.

During the summer of 1919, the Federation of American Motorcyclists disbanded, leaving motorcyclists in a bad way. Specialized journalists, such as Leslie Dick Richards, in *Motorcyclist and Bicyclist Illustrated,* denounced the failure of the federation. To spare the manufacturers and riders of two-wheelers a serious crisis it would have to be completely recreated and reorganized. And so H.P. Parson, the director of the aforementioned newspaper, became the head of the newly formed American Motorcycle Association.

As the politics of war subsided, the company from Milwaukee decided to begin racing again with its original rider, Bill Ottaway, who rehired Red Parkhurst and Otto Walker. Before long people were talking about them and their motorcycles again. At a 100-mile race at Ascot Park in Los Angeles, in front of 25,000 people, on January 4, 1920, the first four places were taken by Harley-Davidson, and the brand from Milwaukee was back among the best.

New records followed, with their eight-valve motorcycle reaching almost 112 mph on Daytona Beach. On February 10, 1920, the very expert and skilful Harley-Davidson race service engineers returned to this beach with riders Fred Ludlow, Red Parkhurst, and Otto Walker to set an official record.

On February 13, 1920, on soft, damp ground, Fred Ludlow reached 103 mph. On the same day, Red Parkhurst reached 110.94 mph. The following day, with dry weather, the record was broken at 112.61 mph, but the International Motorcyclists Federation did not accept this record because certain rules had to be respected: there could not be a tail wind and the time had to be measured in both directions.

Luckily, a few days later, on February 22, 1921, the Company's efforts were rewarded in California, when a Harley-Davidson snatched victory and set a new record on a wooden track with an average speed of 96 mph.

During this period, Ottaway reorganized the support given to Harley-Davidson motorcycles during races, rigorously training all his staff to perform pit stops in only a few seconds. Wheels were replaced, oil and gas tanks filled up (accelerated by a larger orifice), and each rider received a clean pair of goggles – all in less than a minute.

Between the years 1919 and 1921, the Harley-Davidson Motor Company manufactured a single-cylinder model

with valves, the Series W Sport Twin or "SW," driven successfully by Ralph Hepburn. A model of this new racing beast was sent overseas to England, where it was modified by the British rider Freddie Dixon. Subsequently, Dixon won many victories with it on his home turf.

On April 28, 1921, the first motorcycle to reach 160 km/h (100.76 mph) on a circuit was a Harley-Davidson, driven by a manufacturer with the same name, Douglas Davidson, in Brooklands, England.

Despite its success in racing, Harley-Davidson also had to keep improving its commercial range to remain competitive in the touring motorcycle market. In 1926, Harley-Davidson stopped participating in races. Despite the beneficial publicity generated by its victories, the sum of money involved – $200,000 – was simply too high.

***59 bottom** The hill-climb, a race that still exists today, developed around 1915. Until 1920, riders used standard models, modified with rear-wheel chains or belts to improve traction. These metal belts were made from riveted or welded metal parts which were attached to the wheel when the tire was deflated. Inflating the tire then moved the parts into the right position. Screwed U-shape forks tightened belts and rim to avoid any slipping.*
However, from the mid-1920s, machines were designed specifically for hill-climb races. Built with only one gear, they were designed to climb to the top of a steep, uneven hill as fast as possible. Harley-Davidson, Indian, and Excelsior all produced special hill-climb machines.

The 1921 Racer

Perhaps anticipating the rising costs of racing, as early as 1921, the Harley-Davidson Motor Company did not really intend to invest in expensive research in order to manufacture a new, stronger, and more efficient engine designed to win races. At the time, most of the circuits were entirely covered in wood with banked curves, called "board track," or they were simply compacted earth with flat curves, called "dirt track." However, the company did decide to develop a mass-produced "V-Twin" engine with a few modifications to guarantee its durability while making it more efficient. The main modification concerned the cylinder heads. Each cylinder had two intake valves and two cutouts, controlled by means of tappets and rocker arms, driven by a single camshaft housed in the timing case.

This engine was built thanks to the experience and knowledge of a specialist, Harry Ricardo. In the meantime, an English rider, Freddie Dixon, fitted two carburetors on his racing bike with more than satisfactory results. The "Racer" of the early days had only one carburetor though there were three versions of this eight-valve engine. They cannot be confused with each other, for each was different and corresponded to the needs of the rider.

This engine fits inside a frame, nicknamed "Keystone"; a lower element formed by two plates sandwiched the engine block. Don't look for brakes: at that time, brakes were not allowed on racing bikes, because their use was considered dangerous for the other competitors!

The "Racer" was fitted with a primary and secondary chain transmission with an intermediate pinion which adapted the chain tension by means of a cam. This superb bike was reserved for professional drivers, and its prohibitive price ($1,500) was three times greater than that of any other racing bike. That alone was enough to disuade non-professionals.

Since no more than a dozen of this model were built, some of which were sent overseas to distinguish the brand from the competition on European circuits, this has become a very rare collector's piece. Willie G. Davidson owns a superb copy built by Steve Wright. The one shown in these pictures, however, belongs to an Italian collector who is well aware of the value of his treasure, even though the frame is not original.

60-61 The Racer, a Harley-Davidson adapted from the touring model, was fitted with a 61 cubic-inch (1000 cc) 8-valve engine. From early on it stood out, winning races when ridden by Otto Walker. It was the first machine to go faster than 100 miles per hour during a one-mile board track race at Fresno, California, on February 22nd, 1921. On that day, riding this motorcycle, Otto Walker won every competition in which he entered.

Smoky Joe

In 1904, one year after the first Harley Davidson was built in Wisconsin, Joe Petrali was born in San Francisco. Born on February 22, Petrali had an early, boundless admiration for motorcycles. His passion was encouraged by a neighbor called Dewey Houghton, who was a mechanic. Houghton's motorcycle, was a Flanders, an American brand manufactured for the few years between 1911 and 1914. Fitted with a 499 cc single-cylinder engine, its general appearance was similar to that of the "Silent Gray Fellow," which had come out a few years earlier. Dewey Houghton looked after his motorcycle very carefully, and sometimes he would demonstrate the meticulous technique required for tuning the Flanders engine.

Joe Petrali's father gave his son his first motorcycle, an Indian, when Joe was thirteen. This motorcycle had been ridden hard for a number of years on compacted earth tracks. Nonetheless, Joe had already developed a taste for speed and victory, and was determined to get his bike ready for racing. Working with an Indian dealer from California, he prepared well.

In 1918, Joe Petrali won the 500 cc category of the Economy Run, a race where one needed to cover the longest distance using the minimum amount of fuel. On that day, Joe managed to run for 175 miles on less than a gallon, thus claiming the national record and inscribing his name in the history of American racing.

This victory on his very worn-out motorcycle was followed by other successes, and Petrali soon attracted the interest of motorcycle racing patrons, who realized that, despite his tender age, he had the makings of a great rider. When, in August of 1922, "Shrimp" Burns, the official Indian rider, was killed during a race in Toledo, Ohio, Joe Petrali took over. During one long race, a bad fuel mixture caused serious carburation problems in the eight-valve engine. Inspite of this, the young champion finished the race second.

This was an enormous achievement, especially when one considers that he was competing against drivers much older and more experienced than he.

When he was 16, Petrali had left California to move to Kansas City. He worked there as a mechanic for a motorcycle dealer, Albert G. Crocker, who also manufactured his own brand from 1936 to 1941. Crocker took advantage of some of the research and development done by the major brands, using a number of Harley-Davidson and Indian parts in his V-Twin engines. Crocker hired Petrali based on the young man's reputation, and he was not disappointed.

For a board track race in Altoona, Pennsylvania, Indian had guaranteed Petrali that his personal engine would be overhauled, prepared, and refitted on to its original frame in time for the start. But on the day of the race, due to a shipping error, his engine had not arrived. Making the best of a bad situation, Joe borrowed the Harley-Davidson of a rider who had been injured during the tests. They agreed to share the prize money in the event that Petrali won the race. During the race, Petrali was so focused on winning that he didn't notice when another rider had to stop to change his front tire and lost two laps. Seeing the same rider in front of him later in the race, Petrali struggled to pass and went two extra laps, not realizing that he'd already won.

His fame was increasing and in 1921 Harley-Davidson hired him away from Indian and made him their official rider. In his new role, he continued to claim victories in races as varied as the flat track, the board track, and the hill-climb. When Harley-Davidson interrupted its sporting activity from 1926 to 1931, Joe Petrali turned to the Excelsior Supply Company, which manufactured motorcycles in Chicago from 1907 to 1931. With the collaboration of Ignatz Schwinn, Joe Petrali built a series of Excelsior V-Twins called "Super X." During this period, but before his Excelsior was ready, Petrali entered a race in Springfield, Illinois, riding a 350cc single-cylinder Indian. During the race, his friend and rival on the circuits, Eddie Brink, lost control of his bike coming out of a curve, skidded, and slammed into Petrali's Indian. Eddie Brink was killed on the spot; Petrali suffered a massive head injury and a seriously torn lip. Thanks to the intervention of a young doctor and Petrali's own strong constitution, he was soon back on his feet and ready to start racing again.

Four years later, when Petrali was twenty-seven, the managers of the racing group at Harley-Davidson resumed their activities and rehired him for an important series of races. Once again, Joe Petrali became a full-time Harley-Davidson employee, and his experience as both a mechanic and a rider was important to the company.

In 1935, when Petrali was thirty-one, he won thirteen races valid for the national championship. In only one day of racing, he filled his record book – after winning the 1-mile race, he snatched the 5-, 10-, 15-, and 25-mile races on the same circuit in New York. It was on this day that he was given the nickname "Smoky Joe," referring to the cloud of dust he left in his wake.

At that time, the Harley-Davidson Motor Company was just finishing work on a new V-Twin generation, the Series E. Knucklehead

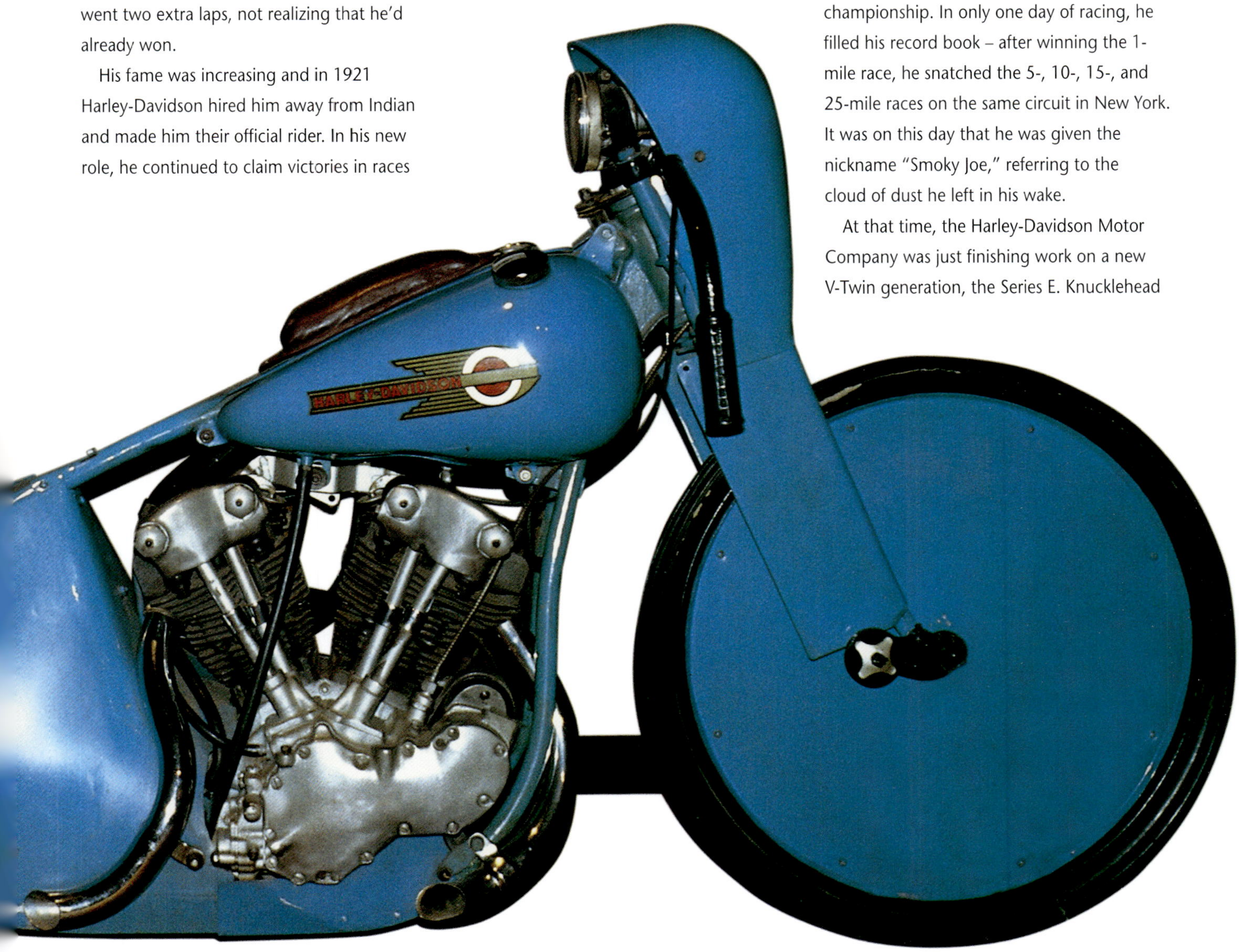

62 Joe Petrali first raced on an Indian. Later, he gained a name for riding Harley-Davidsons and then stole victories for Excelsior, from 1926 to 1931, until that company shut down. In 1931, Harley-Davidson returned to racing and engaged Joe Petrali again. His invincible style stood out on the dirt track, on the board track, and in hill-climbing.

62-63 Joe Petrali gained immortal glory by riding this Harley-Davidson OHV 61, powered by a Knucklehead 1000, at Daytona Beach, Florida. In 1937, he set the 136.183 mph record on the sands of Daytona Beach – a record which was never equalled before such attempts came to an end.

Twin, a 61 cubic-inch (1000 cc) OHV with rocker arms, whose conception was partly due to Smoky Joe. But it had not yet been properly fine-tuned when President William A. Davidson decided to launch it in an attempt to balance the Harley-Davidson accounts.

To advertise this product, Harley-Davidson launched a massive campaign which would make the Knucklehead and Joe Petrali famous throughout the world. Harley-Davidson built an unusual-looking racing motorcycle with a full front wheel, a narrow fork, a tank furnished with a thick pillow on the top, and a small fairing, or more precisely, a fork head, which was fabricated, under William S. Harley's direction, in a tank built especially for the purpose. The engine itself was a Knucklehead with short and direct exhaust pipes. The ignition was by magneto, which had a double carburetor and a highly efficient camshaft. Then came an enormous media blitz, with pictures of the rider and his motorcycle before they left to set a new speed record at Daytona Beach.

On March 13, 1937, surrounded by fans, journalists, and organizers, Joe Petrali was timed at 124 mph. He could have done better, but certain aspects of the bike worried him, such as the fairing, which made the front lighter at high speed, and the odometer, which impaired visibility.

Within the hour, these elements were eliminated or improved and Petrali set off. He broke the world speed record again, this time with a speed of 136.183 mph, a record which has never been beaten or even equalled on a beach. Other brands in the world tried to beat the record, but without success. There is still so much talk about it today that the speed is strictly limited on Daytona Beach, making it impossible to attempt another record on this sandy track.

A year after this performance, Petrali once again avoided serious injury in an otherwise devastating accident, and he finally decided

to put an end to his career as a rider, at the age of thirty-four. He went back to manufacturing race cars for a few years, then he began working for the millionaire Howard Hughes on the construction of his legendary plane "The Spruce Goose." Smoky Joe passed away of natural causes at the age of sixty-eight. An integral part of the Harley-Davidson legend and American motorcycling history, he cannot be separated from the history of world records.

64 Joe Petrali proved to be an extraordinarily talented rider, with a good sense for mechanics that helped him capitalize on innovative systems which his competitors did not dare try. His distinctive riding technique, effective in any type of race, won him a considerable number of victories.

64-65 When Joe Petrali broke a record at Daytona Beach, his motorcycle, fitted with a Knucklehead engine, appeared like this – without its fairing. The fairing, shown during press conferences and used during his first attempt, proved dangerous, since it unbalanced the motorcycle.

Red Parkhurst

The first motorcycle races took place on oval tracks covered in pine, similar to the flooring used in houses. Drivers reached high speeds and sometimes lost control, killing themselves in accidents that also caused injuries to the rows of spectators. Many amateur drivers competed on this deadly surface. One among them, a Harley rider, regularly stood out. Leslie Parkhurst, a native of Colorado, was born to win, and because of his ginger-colored hair, was given the nickname "Red."

The racing group at Harley-Davidson understood that hiring this young rider as an official member of the team could only bring further success, helping to consolidate Harley's already well-established reputation. Red Parkhurst agreed to race for Harley-Davidson on normal tracks, and he added significantly to the numerous victories of the Harley-Davidson Motor Company – until he suffered a serious concussion. Luckily, there were no long-term effects, and the great racer recovered.

When the beginning of the Great War put a sudden stop to the races, at least for brands such as Harley-Davidson, Indian, Excelsior, and a few others, the drivers had to wait for better days and to be patient, often taking jobs that had nothing to do with their passion. Leslie Parkhurst worked for a coal mining company located in the same city as Harley-Davidson until 1919. After the war, he did not want to resume racing but changed his mind after a discussion with Bill Ottaway. He had his new start at a race held in Portland, Oregon, with new and much more powerful bikes. Red Parkhurst remained a leading figure in the company and a top racer. He added the 5-, 10-, and 25-mile races to his records in successive stages.

One of this great rider's qualities was his faithfulness to the brand. Even though he received many offers, some of which were very appealing, he remained with Harley-Davidson, and his loyalty won him the respect of many factory managers and race officials.

Leslie "Red" Parkhurst left racing very late, and was hired by Firestone – a solution that allowed him to stay in touch with motor sports, especially the two-wheeled variety.

66 Red Parkhurst, whose official career began when he attracted the attention of the brand-new Harley-Davidson racing department in 1914, rapidly started scoring victories for the company. He raced again for Harley-Davidson after World War I and retired as the world's first motorcycle celebrity.

66-67 In 1926, Harley-Davidson introduced a new machine, equipped with a 21.10 cubic-inch (345.73 cc) one-cylinder engine. Referred to as "BA," this motorcycle became known as the "21 single," named for its 21 ci OHV engine. The racing model derived from this machine was the famous "Peashooter," so called because of the sound it produced. Variations on this motorcycle continued to be produced, including adaptations with side and overhead valves.

Harley-Davidson Suffers the Great Depression

History and Implications of the Crisis

The crash of the New York Stock Market in late 1929, which devastated the U.S. economy, obviously had an enormous impact on the world of American motorcycle production. Despite bankers' efforts to curb this crash, its aftereffects were felt until 1933. It caused many bankruptcies and triggered an incredible social crisis, with the number of unemployed in the U.S. rising to 15 million in 1933.

Shortly before the crash, at the beginning of 1929, newly elected President Herbert Hoover had made a speech in which he said, "The American society is about to beat poverty forever." He was referring to the economic indicators at that moment: an industrial growth of 64% between 1919 and 1929, and an economic boom in the building and the automobile sectors. In the United States at that time, there was one car for every five people. Advertising encouraged people to spend more and to purchase luxury items. What is more, between 1923 and 1929, although companies profits grew more than 60%, the average worker's salary increased by only 11%, thus widening dramatically the gap between production and potential consumption.

Hoover had thought the United States possessed enough economic power to overcome its difficulties, but it took Franklin

***68** The 74 cubic-inch (1200 cc) VL Sport Solo model with side valves, which was introduced in 1930, had 15 to 20% more power than earlier models. From 1933 on, it sported more attractive paintwork than plain olive green.*

69 In July 1928, Harley-Davidson introduced its Model D, equipped with a 45 cubic-inch V-Twin engine. Presented as a 1929 model, it enjoyed a long, brilliant career, despite the fact that it was launched just as the Great Depression began.

Delano Roosevelt's election in 1932, and his decision to repeal Prohibition and announce the measures known as "the New Deal," for the situation to improve. He put the budget in order, launched works funded by the Federal government, reduced agricultural production, thus raising prices, and relaunched industry. Thanks to the measures adopted between 1933 and 1934, in 1937 American industrial production managed to regain the level it had been at in 1929. But in 1940 there were still eight million unemployed. Only America's entry into World War II, in late 1941, would eventually bring full recovery to the U.S. economy.

Despite these economic hardships, in 1929, the Harley-Davidson company produced its new 45 cubic-inch, side-valve engine, which

they called the "Flathead." It was both powerful and versatile, and the company hoped that it would become its new workhorse. The Flathead was an immediate success and remained so for many years: not only did it manage to go through the Great Depression and World War II, it equipped the Servi-Cars until 1974. In 1930, the company introduced the VL74, whose creation set a new standard for touring motorcycles. It still had side valves but the engine was enlarged to 1200 cc. The company also produced a 30 cubic-inch (500 cc) single-cylinder engine manufactured until 1935.

In 1932, a three-wheel utility vehicle, which would become very popular with the police and other civic departments, appeared in the catalog, listed at $450. It was, of course, the famous Series G Servi-Car, which, with its V-Twin 45 cubic-inch (750 cc) engine, would be manufactured for more than forty years.

Nevertheless, despite these successes, Harley-Davidson continued to experience the same economic crisis as the rest of the country. In the same year the Servi-Car was introduced, Harley-Davidson registered a very sharp drop in production, manufacturing only 7,217 motorcycles. The drop continued in 1933, with 3,703 bikes coming out of the Milwaukee factory. Luckily, the level of production recovered in 1934, and rose to 11,212 motorcycles.

***70 top and center left** William H. Davidson, who was to be president of the Harley-Davidson Motor Company from 1942 to 1971, winning the Jack Pine Enduro in 1930.*

***70 center right** To boost sales, Harley-Davidson offered a new range of colors and ornamentation from 1923 onwards.*

***70 bottom left** The Servi-Car, equipped with the 750 cc (45 cubic-inch) side-valve engine, was first listed in the Harley-Davidson catalog in 1932. It soon became popular for law enforcement and other professional purposes.*

***70 bottom right** The Harley-Davidson line was already amazingly varied and the company was good at advertising this fact. It offered vehicles for any use, bragging that even women could easily ride a Harley-Davidson.*

In that period, motorcycle manufacturers experienced overwhelming difficulties and nearly all disappeared from the market, because leisure activities had taken a back seat to professional migrations and the public was no longer as interested in the motorcycle. The car also absorbed a considerable share of motorcycle sales.

It was a period of scrimping and saving. Sales dropped dramatically, and some employees worked only two or three days a week. This reduction in working time helped to avoid massive layoffs.

Given the economic climate, Harley-Davidson had to use a great deal of ingenuity and imagination to sell its new models, not to mention the rest of its line. The manufacturing methods also improved considerably and nothing was left to chance. Each promising idea was thoroughly analyzed before being implemented. Despite this cautious approach, a number of technical improvements were made: aluminum pistons, interchangeable wheels, a better electrical layout, thanks to a new ignition coil and, finally, warning lights, to help riders check the bike's operation.

In 1933, Harley-Davidson decided to brighten up its motorcycles. It gave them a cost-effective but attractive new look simply by varying the colors and decorations. The khaki color of the mudguards, the tanks, the odometers and the frames was replaced by variegated colors, which were further brightened up by new logos.

But the motorcycle manufacturers' situation became increasingly worrisome; in 1933, overall motorcycle production, all brands included, dropped to 6,000 units. Harley-Davidson did its best to avoid bankruptcy.

The American automobile industry and its suppliers registered an 80% drop in production, and they too found themselves

71 This new type of tank decor appeared in 1934, but it was only available for two years, since Harley-Davidson was becoming aware that variations in color and logo could boost sales.

in a very delicate situation. In 1933, one estimate reported 100,000 registered bankruptcies. With exports reduced to nothing and unprecedented price cuts, the United States was experiencing the worst economic catastrophe in its history.

Gradually, the Harley-Davidson Motor Company emerged from the chaos. In 1936, the company offered a few new models. That year also introduced of the famous "Knucklehead" engine on a bike in the E series, EL 61 OHV; it was available with either normal or high compression. It was the first Harley-Davidson to have its dashboard integrated into its tank, as it still is today on most of the brand's models. It was also the first Harley-Davidson to have a real oil circuit with a separate sump, as we know it today, though admittedly the technology was not perfect on those early models. It also had a four-ratio gear box, which was a technical improvement that left its competition at

Indian in the dust. This machine offered more power by doing away with the side-valve system, and it met with the success it deserved.

The FL Special Sport Solo, a 1200 cc model which had a 74 cubic-inch, high-compression Knucklehead OHV V-Twin engine, was put on the market in 1941, originally to meet the requirements of various police forces. This model's successful career, together with that of the 1000 cc model, the EL 61, continued until the appearance of the Panhead engine in 1948. But, before that event, the Harley-Davidson Company would play an active part in the Second World War.

***72** The Knucklehead was presented during the summer in 1936. It had an overhead valve, 1000 cc engine that provided more power than the Flathead it replaced. In 1937, Harley-Davidson adopted the technique of fixing the dashboard on the top of the tank – a practice still followed today.*

***73** A vintage picture, taken to promote the Knucklehead. Here (from left to right) Arthur Davidson, Walter Davidson, William S. Harley, and William A. Davidson are meant to be inspecting the very first Knucklehead 61 coming off the 1936 assembly line.*

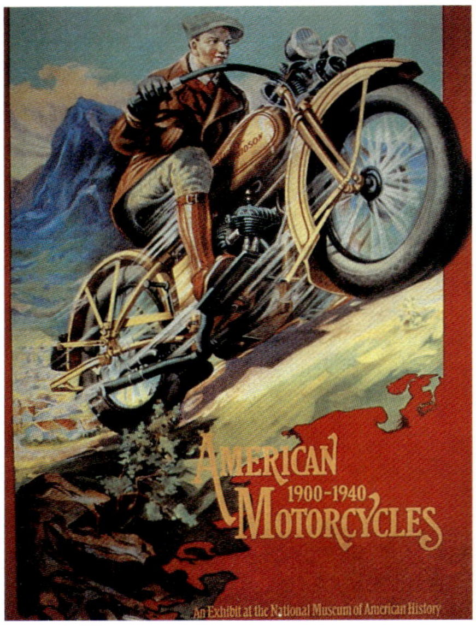

74 top The Harley-Davidson name gradually became synonymous with motorcycling in the U.S. By the early 1940s, there were only two big manufacturers competing with each other: Harley-Davidson and Indian.

74 center and bottom Those Harley-Davidsons that were outfitted with the Flathead, the famous side-valve engine, helped to build the myth of Harley-Davidson as a legendary motorcycle. After World War II, army motorcycles, sold by Americans in Europe or left as contributions to reconstruction, rapidly developed a strong Harley-Davidson following among overseas motorcycle fans.

75 In the United States, Harley-Davidsons fitted with side-valve engines were real collectors' pieces. Army motorcycles proved to be rarer in the U.S. than in Europe after World War II. Nevertheless, Harley-Davidson always took its place at important shows, represented by the big 74 cubic-inch (1200 cc) and 80 cubic-inch (1340 cc) side-valve engine models.

***76 left** Because of Harley-Davidson's military commitment during World War II, all civilian motorcycles produced between 1941 and 1946 were identical. Technological and aesthetic detailing came second to the company's commitment to the war effort.*

***76-77 top** Once indispensable among police and other professionals, Servi-Cars today are still sought after, both as collectors' pieces and also as working vehicles. Often they are renewed and decorated according to their owners' tastes.*

77 bottom left From 1936 on, Harley-Davidson offered a series of motorcycle accessories that could be attached right on the assembly line. A purchaser could pick up a machine at the dealer's, already personalized to his taste.

77 bottom right This exploded view of the Knucklehead engine shows its operations, above all its cylinder head lubrication. Early Knuckleheads had a few lubrication problems, which were rapidly solved on the models in the following years.

A Great War Followed

In September 1939, twenty-one years after the end of World War I, Europe experienced the beginning of its second conflict. President Roosevelt asked the American people and the Congress to be ready to intervene, if necessary. On November 4, 1941, America entered the fray. The United States lifted the arms embargo and began producing an exceptional amount of military equipment bound for the Allies. Annual costs for war equipment climbed from $17 to $50 billion, and the United States became a veritable arsenal for the Allies. In the space of four years, the U.S. produced 95,000 airplanes, 3 million tons of warships, 80,000 artillery guns, and nearly 90,000 Harley-Davidsons.

The WLA 45 models, a military version of the WL 45, were the most widely manufactured Harley-Davidson motorcycles used in the conflict. Another type of motorcycle, the XA, with a Flat-Twin engine modeled on that of the German BMW, appeared as a military vehicle. It was able to compete with the European machines, but its career did not extend beyond the war.

Times of upheaval are often times of innovation as well; during this period, Harley-Davidson did not only manufacture motorcycles. Willys, the famous manufacturer of another American legend, the Jeep, planned to build 5,000 small, cross-country cars – a sort of mini-Jeep – that would be powered by 750 cc XA engines. These Jeeps were to be air-lifted to Europe, but the war came to an end before completion of this important order, and the production of X engines did not exceed 1,000 units.

At the conclusion of the Second World War, Harley-Davidson revealed a military secret

78 A World War II armed division. These soldiers are equipped with 750 cc WLAs decorated with division colors and provided with the standard equipment, including holsters attached to the motorcycle forks.

79 Harley-Davidsons could be found in all facets of armed conflict. Many pictures showed how riders could use their motorcycles as shields, although a tank full of fuel and oil did not provide the ideal protection.

***80** Motorcycles were used for escorting convoys, as rapid liaison, and by the military police, but they were rarely found at the front or on reconnaissance. Jeeps stole the limelight in this phase of conflict.*

regarding another military vehicle: the company had supplied certain Canadian mini-tanks with two 61 cubic-inch Knucklehead V-Twin engines. However, this project remained experimental and the machines never saw full military service.

In recognition of the Harley-Davidson Motor Company's contribution to the U.S. war effort, it was awarded the U.S. Army-Navy Production Award, known as the "E" award for excellence.

With the war over and military requisitioning at an end, civilians bought up the stock of available WLAs: nearly 15,000 Harley-Davidson WLA 45s were sold at a unit price approaching the original price of $500. During the '50s, some of WLAs were still sold for as little as $80. Thanks to the strength and reliability demonstrated by these war-time motorcycles, the Harley-Davidson reputation was reinforced in Europe, and as a result, the company renewed efforts to establish its brand.

Despite its war-time success, it would be eighteen years before Harley-Davidson won another Army contract. The order was for a model called the XLA, which was a Sportster fitted with rigid fiberglass bags and a windshield. Although this was the last true Harley-Davidson to be used by the Army, on October 10, 1987, Harley-Davidson bought the manufacturing rights for single-cylinder 350, 500, and 560 cc motorcycles from the Armstrong Company. These were real cross-country models, which were used by security patrols and escorts, communication and reconnaissance services. The vehicles, fitted with Rotax engines, were never sold to the public.

80-81 Sidecar prototypes were built for the Army but this type of vehicle was rarely used in the Second World War, as Jeeps proved more efficient and agile on all types of terrain.

81 bottom The President of Harley-Davidson Motor Company, William H. Davidson, proudly accepts, on behalf of all the company employees, the Army and Navy "E" Trophy, as a reward for the war effort.

"WLA" and "XA": Two Reference Points

82 In addition to the WLA model, Harley-Davidson supplied the Army with a descriptive motorcycle manual and a troubleshooting guide in case of breakdown.

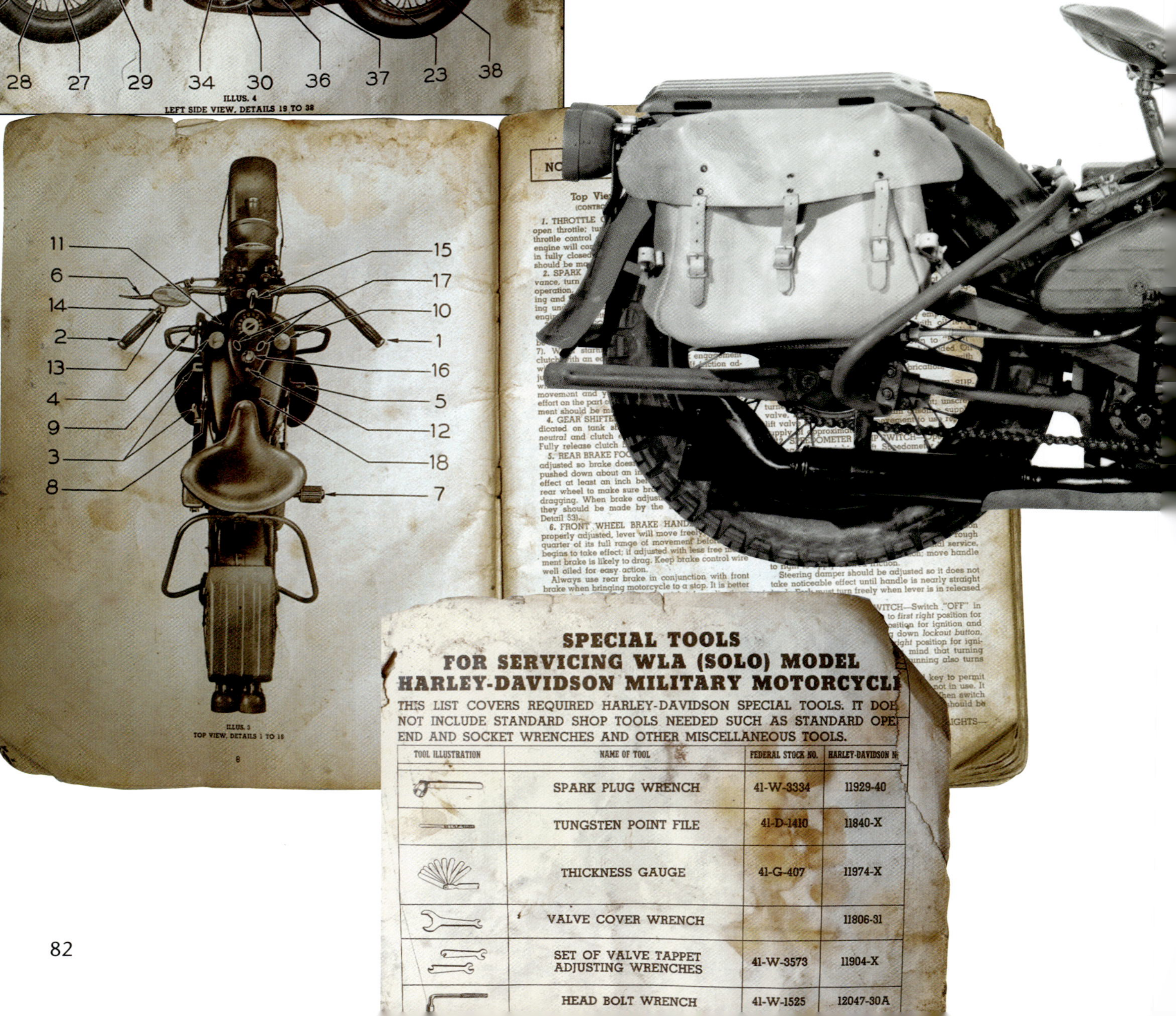

SPECIAL TOOLS
FOR SERVICING WLA (SOLO) MODEL
HARLEY-DAVIDSON MILITARY MOTORCYCL

THIS LIST COVERS REQUIRED HARLEY-DAVIDSON SPECIAL TOOLS. IT DOE NOT INCLUDE STANDARD SHOP TOOLS NEEDED SUCH AS STANDARD OPE END AND SOCKET WRENCHES AND OTHER MISCELLANEOUS TOOLS.

TOOL ILLUSTRATION	NAME OF TOOL	FEDERAL STOCK NO.	HARLEY-DAVIDSON N
	SPARK PLUG WRENCH	41-W-3334	11929-40
	TUNGSTEN POINT FILE	41-D-1410	11840-X
	THICKNESS GAUGE	41-G-407	11974-X
	VALVE COVER WRENCH		11806-31
	SET OF VALVE TAPPET ADJUSTING WRENCHES	41-W-3573	11904-X
	HEAD BOLT WRENCH	41-W-1525	12047-30A

***82-83 and 83** Used on all types of terrain, the WLAs proved to be remarkably versatile. About 88,000 such motorcycles were produced for the Army.*

As evidenced by the "E" award the company was given at the end of the war, like many American companies, Harley-Davidson played an important role in the conflict. For a second time, William S. Harley had the responsibility of adapting Harley-Davidsons to the harsh conditions of war. From 1939 on, Harley-Davidson trained its mechanics to be ready for all eventualities. Their duty was to guard against potential failures and to maintain the WLAs on the battlefield. Every four weeks, almost fifty trained mechanics completed the training course at the school located at the company premises on Juneau Avenue. The U.S. Department of Defense signed a contract with the Harley-Davidson Motor Company and with the Indian Company for the manufacture, almost exclusively for the war, of about 88,000 WLA 45s, a military version of the Harley-Davidson WL 45. By modernizing its production techniques, Harley-Davidson was able to supply this large number of motorcycles. The Harley-Davidson Motor Company decided to expand its production area and capacity by renting supplementary space in the area near its factory. Moreover, it organized its workers into shifts so that they worked round the clock. By these means, the company's assembly lines were able to produce enough WLA 45s and WLA 74s, not to mention all the spare parts. In fact, the WLA 45's role in the liberation of Europe won it the nickname "Liberator."

All the tests on these military models were carried out at Fort Knox under the supervision of military personnel and, until his death from a heart attack on September 18, 1943, William S. Harley). The wheels were interchangeable and had an 18-inch diameter. The tires were 4 or 5 inches wide, and the rider was protected from the spray by very large mudguards, fitted at the highest possible level to prevent the mud from blocking the wheels. The large and well-protected air filter enabled the bikes to cross shallow fords. A steel plate was fitted under the lower part of the frame to protect the engine, and the chain was also protected by a large, strong guard. An internal lubrication system allowed cooling during long periods of use at low speed. A rugged metal luggage rack was provided to house a two-way radio weighing 40 pounds. On the right front, a gun holder was fixed for a Thompson machine gun or a light automatic carbine. On the left part of the fork, a metal box acted as counterbalance and contained magazines and ammunitions.

Certain WLAs were also equipped with a hand-cranked siren, operated by a shaft that rubbed against the rear tire. This surplus weight prevented these WLAs from going faster than 50 mph, and its acceleration was equally reduced. The motorcycle behaved well on the battlefields, though, not least because it was easy to maintain. Harley-Davidson also manufactured 20,000 WLAs for the Canadian army under the designation WLC and also ELC. In the Canadian versions, a few elements changed position as compared to the American and European models: the gear lever and foot clutch were moved to the right, while the brakes were moved to the left, on the handlebar.

The U.S. Army was so impressed by the performance of the German BMWs and the Zundapps that, in the spring of 1942, they asked Indian and Harley-Davidson to come up with a similar motorcycle with equivalent performance capability. The XA, which could out-perform the WLA, was suggested as its replacement.

84 Since the WLA was the Harley-Davidson model most often ordered by the Army, it was produced in large quantities. By the end of World War II, it was possible to find many WLA spare parts in Europe.

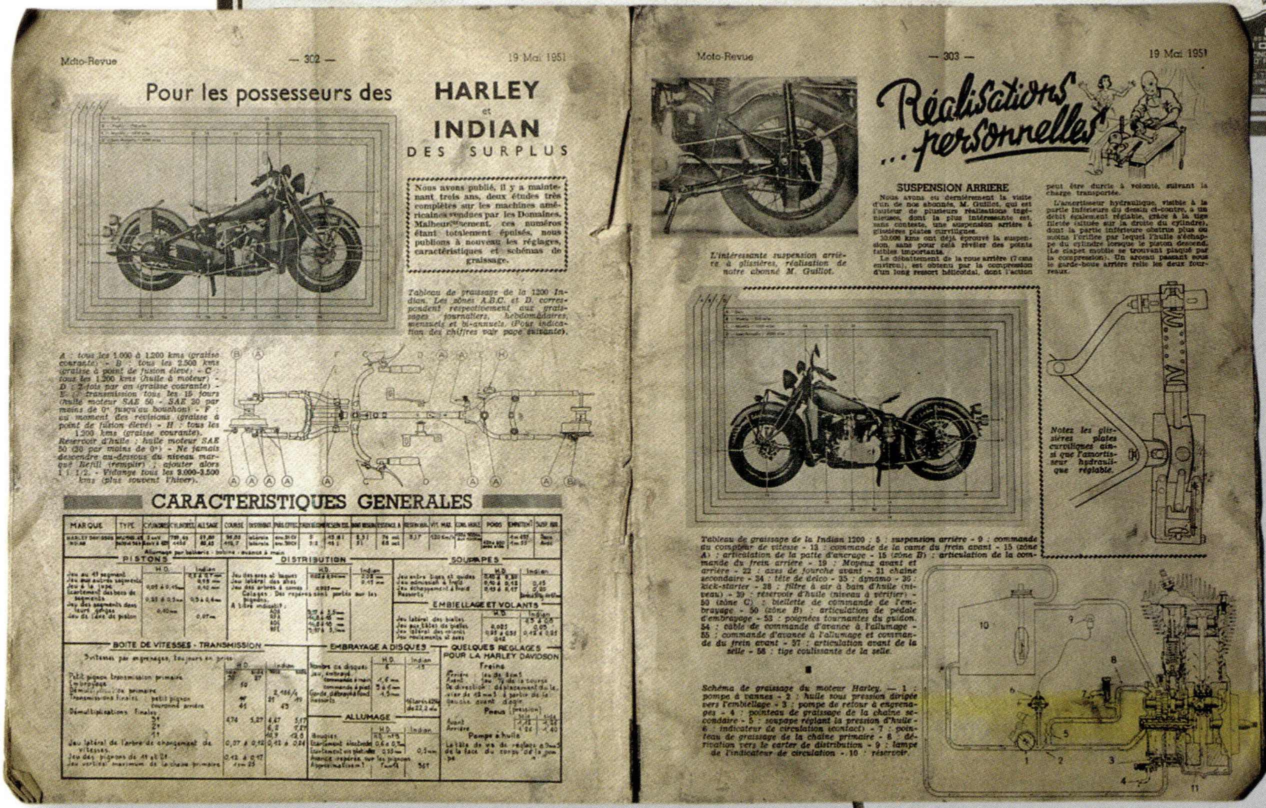

85 While the European motorcycle industry was devastated by the conflicts of World War II, WLAs with spare parts for 30,000 had been shipped to the European territory. So several countries, including France, chose the Harley-Davidson WLA as their army motorcycle.

86-87 Today, many WL and WLA models appear in exhibitions, shows, and meetings in the United States. Some have been modified, while others have been restored exactly to their original state, as if they were coming out from the assembly lines of the Milwaukee factory. Sometimes their owners dress in the military style of the time, parading their motorcycles in a spirit dating back to the early 1940s.

10
20
30
40
50
60
70
80
90
100
110
120
MOTOR
HARLEY DAVIDSON
CYCLES

The XA increased the stock of military equipment, and its engine was inspired significantly, not to say entirely, by the BMW flat two-cylinder of that time. Most details were copied by simply transforming the European measurements into inches. As previously mentioned, in the end, only 1,000 models of this motorcycle were built. Fitted with a fork bordered by hydraulic shock absorbers, their cost to the U.S. Army was less than $1,000 per unit. After the war, civilian customers were eager to own this motorcycle, which appeared revolutionary, but only 200 XAs were turned into civilian models and preserved by collectors. Thus, the XA did not manage to depose the 45 degree V-Twin, which had prevailed since 1909 and which was already available in different versions.

***88 top** During World War II, Harley-Davidson's training school for mechanics enrolled only soldiers, teaching them to solve motorcycle problems in the field.*

***88-89 and 89** American soldiers were impressed by the efficiency of German motorcycles, especially their reduced maintenance requirements despite intensive use on difficult terrain, such as sand. As a result, they asked for similar motorcycles. Thus the XA, a copy of the BMW, appeared. About a thousand motorcycles of this type were built.*

Return to Civilization

End of a Conflict and Economic Competition

Before Harley-Davidson could resume large production volumes, they had to rebuild what the war had damaged. The Harley-Davidson Company encouraged its agents to reestablish the dealer network again, but the dealers needed to be persuaded that the company was still efficient and serious about business, that the war was really finished, and that the models currently being manufactured would still attract the public. The commercial and entertainment press and *The Enthusiast*, the already celebrated company magazine, spread the Harley-Davidson message far and wide. Beyond reassuring dealers, this message was also aimed at once again attracting civilian customers, who had been relatively neglected during the war. Meanwhile, a new fashion, one on two small wheels, invaded the United States and became especially popular among students seeking a cheap means of transportation. A company from Lincoln, Nebraska, called Cushman, which belonged to the Johnson-Evinrude group, began marketing the scooter in 1944. After the war, the 4.8 horsepower Cushman sold for about $150, and to help them to rebalance their accounts, many Harley-Davidson dealers started selling this successful product.

Harley-Davidson was less than enthusiastic about this situation. In the spring of 1947, the managers in Milwaukee threatened to allow only official dealers and not franchise holders to sell their motorcycles. Thus they forced dealers to choose to represent either Harley-Davidson or Cushman. This conflict was heightened at the end of 1947, when Harley-Davidson began offering a model that broke somewhat with the Harley-Davidson mold. It was a small motorcycle fitted with a 124 cc two-stroke single-cylinder engine. Its appearance was faithful to the Harley-Davidson spirit and, like the Hydra-Glide, its tank was graced by an entirely chrome-plated, water-drop-shaped logo with Harley-Davidson carved in black. This motorcycle marked the start of a production of small-displacement motorcycles, which charmed a new group of motorcyclists, who found these light bikes to their taste: women.

By diversifying their production and improving their Milwaukee factory, which had again become too small to work efficiently, Harley-Davidson gradually resumed a normal production pace. In 1945, the Harley-Davidson Motor Company bought some buildings located a few minutes from Milwaukee for $1.5 million. The 260,000-square-foot premises were located on Capitol Drive in Wauwatosa; today, they house production of the Harley-Davidson V-Twin engine line. In 1949, they enabled the company to embark on the manufacturing of new models, including the Hydra-Glide. The Hydra-Glide had the first hydraulic fork and was driven by the new generation of Panhead V-Twin, fitted with aluminum heads and adjustable hydraulic tappets. This post-war model is considered the ancestor of the Electra Glide and, today, Harley-Davidson is keeping the Hydra-Glide spirit alive with the "Heritage Softail Classic" model.

***90 top** In 1948, in order to gain greater market share, Harley-Davidson built a small, 125 cc three-stroke one-cylinder motorcycle, the S Lightweight, inspired by the German DKW.*

***90 bottom** The last Knucklehead, with a newly designed tank logo and a new dashboard, was built in 1947. From 1941 onwards, the Knucklehead had been available in either 74 cubic-inch (1200 cc) or 61 cubic-inch (1000 cc) versions, which enjoyed parallel careers.*

91 top The Servi-Car, still produced during the war, went through an evolution of improvements, such as this 1948 model. Practical accessories were added to answer the needs of those who used it for business.

91 center Before World War II, during the war years, and immediately after the end of hostilities, Harley-Davidson continued producing side-valve engines. Their civilian versions were available with various capacities: 45 cubic inch (750 cc), 74 cubic inch (1200 cc), and 80 cubic inch (1340 cc). All three models are still very popular today, and are often exhibited at shows. Their excellent reliability allows them to be customized and used by their owners decades after their original manufacture.

91 bottom Since war production was a priority for Harley-Davidson, the 1941-1946 civilian line did not develop and for five years had practically the same look. This type of tank, with this logo, appeared on all those civilian motorcycles, whether they were equipped with a Flathead or a Knucklehead engine.

92 top and left **From 1949 onwards, Harley-Davidson's new flagship engine, the Panhead, was fixed on a frame fitted with a telescoping fork, replacing the old Springer.**

But more difficulties were ahead. Walter Davidson died in 1942, and another founding member of the brand, William S. Harley, passed away the following year. In February 1942, the company presidency was assumed by William Herbert Davidson, William A. Davidson's son, who had started working for the company in 1928 as a worker. Because of his background, he was familiar with all the stages of production and marketing. Moreover, he was an excellent rider who successfully took part in many races for the brand; in 1930, driving a 750 cc Model DLD, he won the "Jack Pine Endurance Race." Despite William Herbert Davidson's involvment, some members of the Davidson family were not attracted by a career in motorcycles, and gradually the people responsible for the good operation of the company changed. Additionally, in 1946, despite the difficulties involved for new importers, foreign motorcycles arrived on the American market.

92-93 and 93 In 1948, the Harley-Davidson Motor Company replaced the Knucklehead engine with the Panhead, fitted with aluminum cylinder heads and hydraulic tappets. Two versions of this engine were immediately available: the 61 cubic-inch (1000 cc) and the 74 cubic-inch (1200 cc) models.

The British Invasion

94-95 *The Hydra-Glide had accessories and equipment to counter the English motorcycle invasion, but it proved to be heavy and inefficient compared to its rivals. In 1955, a new model which had higher compression and a more efficient camshaft was introduced: the FLH Super Sport.*

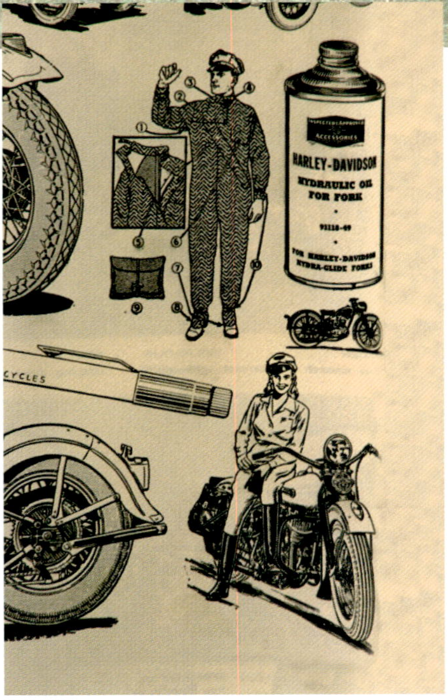

Norton, Royal Enfield, BSA, and Ariel are all British brands which have faded into history, although collectors and enthusiasts continue to find pleasure in riding them. These British motorcycles, which started arriving in the United States after the end of the war, were the first serious competitors from abroad. Despite lines that were less majestic than Harley-Davidson's, they appealed to American motorcycle riders for their sporty appearance and for certain technical improvements, such as the foot gear shift and the hand clutch – techniques which were later adopted by the American brand. They were of interest to the motorcycle dealers, too, because favorable tax laws encouraged their importation.

In 1946, some Ariels, a British brand which first went into production in 1902, a year before Harley-Davidson, started to show up on American highways. This brand had begun by manufacturing tricycles powered by De Dion Bouton's engines. During the Second World War, Ariel also made motorcycles for the army, using 347 cc engines with rocker arms. Afterwards, Ariel and BSA merged. BSA, which had begun manufacturing motorcycles in 1906, continued to produce parts for both brands until 1970, and produced BSA motorcycles until 1973. Triumph, which started off in the same year as Harley-Davidson, was the premier British manufacturer for a long time. Nearly 10,000 of these British motorcycles were sold on the American market in 1947.

The A.M.A. responded to this British threat by enacting regulations favoring motorcycles made in the United States. Harley-Davidson also responded by releasing, in 1949, the Hydra-Glide. This model boasted many innovations and was destined to have a brilliant career. Production resumed its ascent again with 16,222 bikes in 1946, 20,115 in 1947, and 29,612 in 1948. After 1948, it settled on an average of 15,000 to 20,000 motorcycles per year. Profits were also consolidated by the sale of clothing and of accessories that allowed riders to personalize their bikes. In order to develop this commercial sector, Harley-Davidson organized races where the best-equipped and most original motorcycles received awards. Over the years, the motorcycles presented at these events came to look like real works of art, reflecting their owners' unique personalities and abilities.

Harley-Davidson's Response

In 1948, a year after the Harley-Davidson Company bought the Wauwatosa factory, it presented its new 125 cc light single-cylinder engine, the "S" Model, which was produced until 1952. It could reach 49 mph and was an economical model for those who just wanted to cover short distances, and do so at very low cost. Also in 1948, the Panhead was launched to replace the Knucklehead. That same year, at a large meeting of American and foreign dealers, the company's production manager, Gordon Davidson, announced that he hoped to increase Harley-Davidson production. Production in 1948 did grow to – 29,612 motorcycles – but the following year it dropped back to 23,861. The downward trend continued into the new decade, dipping to 17,168 in 1950, then reaching a low point in 1955, when the company produced just 10,686 motorcycles. Despite these problems, the Harley-Davidson Motor Company launched a number of 21 and 30.5 cubic-inch light motorcycles to try to stop foreign brand penetration; but fighting on both fronts, it also reaffirmed its commitment to the large V-Twin models, which were, in many ways, the ideal machines for this big country.

Even so, by 1950, 40% of the motorcycles registered in the United States were foreign brands. William H. Davidson tried everything he could to convince his dealers to double their efforts to promote the American bikes. In particular he tried to reinforce that Harleys alone were suitable to the weather and geography of the country. The company also made a big effort to prove that the new S model, derived from the German model "DKW," was a strong machine in keeping with the Harley-Davidson tradition. Harley-Davidson managed to deliver about 4,708 of this small-displacement motorcycle in the first year of its production.

After the war, European countries needed to relaunch their economies. Building on the success of their smaller bikes, British manufacturers then began building bigger bikes to compete directly against the large V-Twin, in terms of both performance and reliability. As a result, in 1949, Triumph launched its "Thunderbird," with a displacement of 650 cc, and Norton brought out a 500 cc. But these motorcycles failed to live up to expectations and never really threatened the great American tourer.

Confronted with the development of foreign brands, the Harley-Davidson Motor Company realized that it would once again have to fight to save itself.

96 The Panhead engine was produced in the 61 cubic-inch (1000 cc) and the 74 cubic-inch (1200 cc) versions. In 1952, the smaller version was discontinued in favor of the larger, which was more efficient and met with more commercial success.

96-97 The Harley-Davidson Hydra-Glides, equipped with Panhead engines, were regularly improved by technical developments and modified to meet customer tastes. In 1952, the FL Sport Solo included a modern gear-change system with a foot selector lever and a handlebar clutch. For traditionalists, Harley-Davidson also manufactured a model with the earlier gear-change system.

Unfortunately, misfortunes continued. There were high-profile defeats during races like the Daytona Beach 200-mile race, where Norton entered four riders against the other British and American brands. Harley-Davidson entered its "WR45." Norton took the victory and Harley-Davidson was forced to face the fact that its motorcycles were no longer really competitive. On December 30, 1950, tragedy struck: Arthur Davidson and his wife were killed in a car accident. In that year, production decreased by 6,693 motorcycles. It is possible that this further drop was partly due to the implementation of the Marshall Plan, which was intended to jump-start European economies by granting them important financial help.

During this period, competition worried Harley-Davidson dealers, who asked the company for models equal in weight and power to those upsetting the balance of their trade, especially since they were forced to sell and repair only the brand from Milwaukee. Harley-Davidson's inflexibility ended up sitting badly with its distributors. The dealers figured that they had to live; they believed that if Harley-Davidson refused to adapt to the new market, its obstinacy would bring it to a dead end. Despite the ban, certain dealers began to sell other brands, such as Triumph, for which there was then a large demand. It is worth pointing out that in 1954 Marlon Brando would ride a Triumph in *The Wild One,* a film based on the infamous incident in which bikers ran amok in a small California town.

In 1952, in response to foreign attacks and to customer pressure, Harley-Davidson launched the 45 cubic-inch Series K, which replaced the W series. It had a sprung frame, rear shock absorbers, and a hydraulic front fork. Over the next year, it became available in several versions, such as the "KK" and the "KR," with a magneto and upswept exhausts for heavy motorcycle races; it was targeted to young sports customers. The model was adopted in 1956 by none other than Elvis Presley, who chose a "KH" (which was not yet the Sportster). The May 1956 cover of the Harley-Davidson magazine *The Enthusiast* featured the King on his first red and white Harley-Davidson.

98 top In May 1956, on the cover of **The Enthusiast**, *a new rock 'n' roll star named Elvis Presley appeared astride a Harley-Davidson KH.*

98 bottom Series K replaced the W series from 1952 onwards. Until 1954 it had a 45 cubic-inch (750 cc) capacity, then the KH was fitted with a 55 cubic-inch (990 cc) engine Flathead, which it had until 1956.

99 This Model K Sport Model, with its 45 cubic-inch engine, appeared in the 1952 Harley-Davidson catalog. 1952 and 1953 proved to be momentous years for Harley-Davidson, not least Marlon Brando rode a Triumph in **The Wild One**, *a film inspired by the time when bikers terrorized the small town of Hollister, California, in 1947, thus imbuing the British bike with a roguish glamour.*

100-101 Bikes with Panhead engines are often some of the best customized machines at shows. The engines are often restored to their original state or decorated with chrome, though always in a manner which respects the spirit of the time.

MOTOR
HARLEY DAVIDSON
CYCLES

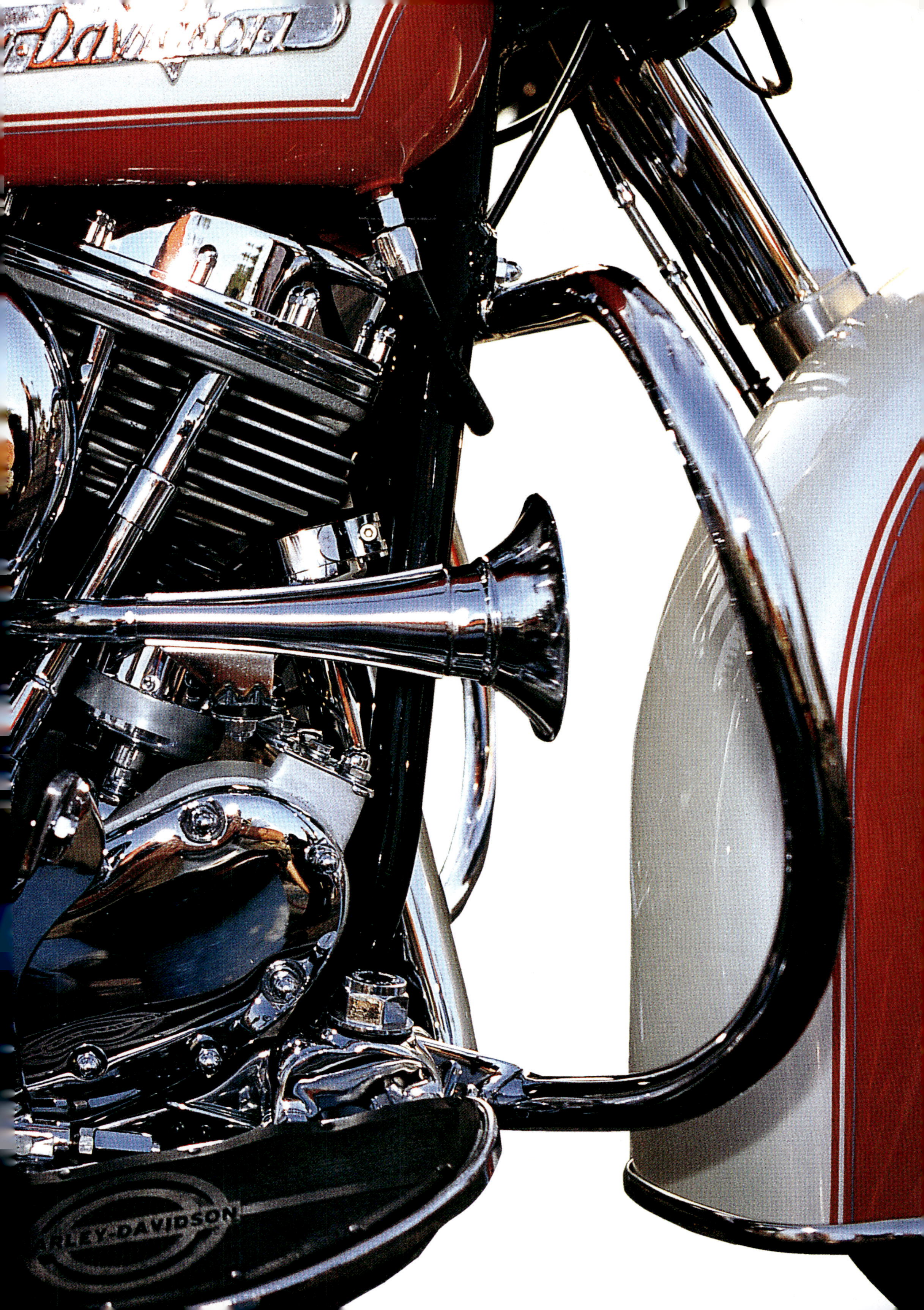
Davidson
ARLEY-DAVIDSON

Creators Are Not Immortal: The Succession

By 1942, the Davidsons and William Harley had constructed a solid enterprise, and since the production of their first engine at the beginning of the century, the sacred brand form the Milwaukee had been a constant presence on roads around the world. During the preceeding forty years, many members of the families had worked for the company, to ensure strength and longevity. But now, foreign competition was not the only problem the company faced.

On February 7, 1942, Walter C. Davidson, Sr., president of the Harley-Davidson Motor Company, died at the age of 66; his death raised questions within the family group concerning the succession. It was not easy to determine the most suitable person to represent such a well-established company. Walter Davidson left his three children, Gordon, Walter Jr., and Robert; his brother, Arthur Davidson; and his two sisters, along with his wife, Emma. Since 1907, over the thirty-five years of his presidency, he had come to personify the Harley-Davidson brand so much that it was difficult to entrust this responsibility to anyone else. Arthur Davidson, the founder, was 63 when the search for a president was taking place; however, his health was not good, and as mentioned previously, he and his wife would soon die in a tragic automobile accident. Although this did not affect the process of choosing an immediate successor to Walter Davidson, it did mean their were fewer family members upon whom the company could rely when the difficulties increased.

The successor had to be diplomatic and flexible and, at the same time, retain the firmness of the first president of the company. Walter Sr. believed that a product conceived honestly and assembled accurately could be sold without qualms to the public. Although two of his sons were involved in the company – Gordon as Sales Manager, and Walter Jr., as Manufacturing Manager – neither of them wished to take on the responsibility of the presidency, and the third and youngest son, Robert, had chosen a career in a completely different sector.

Later, Gordon would leave his job with the company and his brother, Walter, would die of lung cancer, in 1967. In the meantime, William S. Harley decided to leave his position as chief engineer and wished to entrust this

***102** The second and third generations of Davidsons astride their motorcycles: William H. Davidson and his sons, John (at his right) and William G. (at his left), later known as Willie G. Davidson.*

***103 left** Willie G. Davidson, just after winning an endurance race in 1952.*

***103 right** The Davidsons gathered to celebrate the 90th anniversary of the Harley-Davidson Motor Company in Milwaukee in 1993. Mr. and Mrs. Willie G. Davidson (second and third from left) stand with their three children (from left), Karen, Billie G., and Michael. Karen and Billie G. work for the company.*

responsibility to one of his sons, William J. Harley. William J. Harley was happy to accept. He travelled to Italy often during the association with Aermacchi. In 1957, he was appointed vice president of the company, but he died of diabetes in 1971.

His brother, John E. Harley, who was educated at a military college, had the necessary motorcycle experience to train the soldiers of the motorized division at Fort Knox and in Georgia, but he died of cancer in 1976.

The Davidsons were more numerous as Harley-Davidson Motor Company shareholders than the Harleys, with a ratio of 70 to 3 so it is not surprising another Davidson was chosen to take over the role. In the end, William A. Davidson's son, William Herbert Davidson, was deemed to be the most appropriate choice, and he inherited the post of president of the Harley-Davidson Motor Company on February 23, 1942. Holding several diplomas and, in particular, a degree in business, he did not experience great difficulties managing the company. In addition, his uncle, Arthur Davidson, became president of the A.M.A. on January 20, 1944.

Today, despite all the changes over the years, the Davidson name remains within the company, now in the third and fourth generations; and at the end of the century, the person Harley enthusiasts are always eager to meet is William G. Davidson, the most visible symbol of the exciting, epic deeds of the two families. Known as Willie, Davidson once dealt with the styling, the upholstery, the colors, and everything relating to the motorcycle of the year. He is currently responsible for important public relations and promotional activities for the brand, which he carries out in person at the great Harley-Davidson demonstrations and gatherings. His daughter, Karen, and his son, Bill, also work for the Harley-Davidson Company in Milwaukee, while another son lives in New York and works as a painter.

The V-Twins Follow But Do Not Resemble Each Other

***104 top and center** The Sportster stands out in a 1957 issue of* **The Enthusiast**. *It had become Harley-Davidson's strong suit, predicted to counter foreign competition and meet the expectations of customers seeking a lighter, more agile, and more efficient motorcycle than the Big Twin.*

***104 bottom left** Exploded view of a modern Evolution-type Sportster engine.*

***104 bottom** The Sportster promotion campaign targeted those who wanted a motorcycle redolent of youth, performance, and leisure – a really modern machine.*

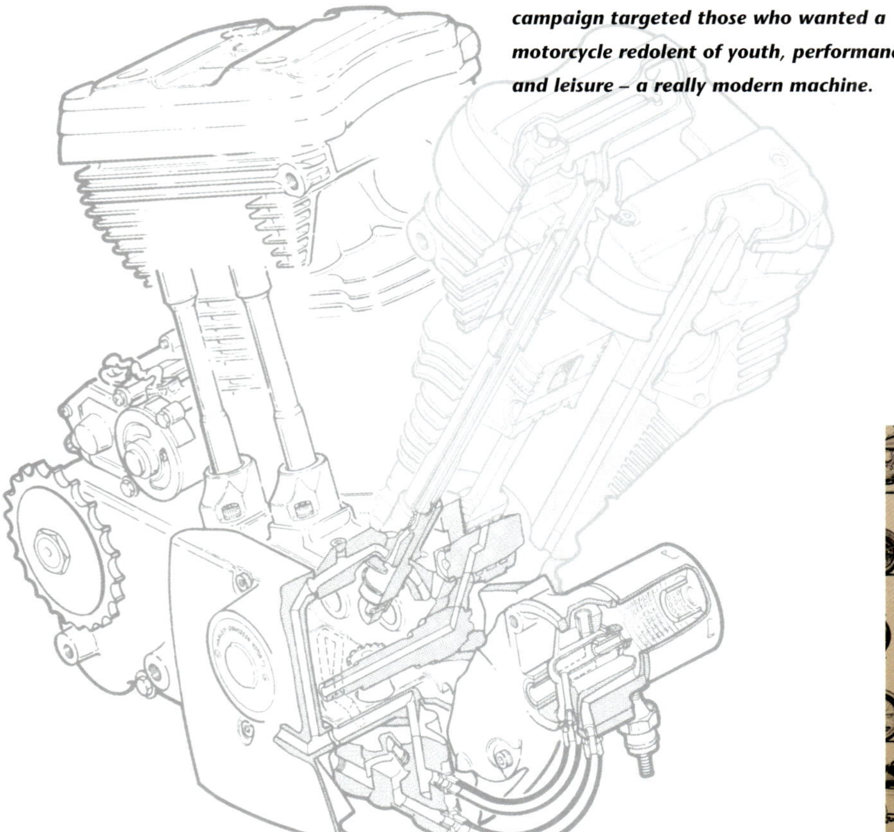

The V-Twins with Panhead engines continued their career and, along with them, a new line was developed based on the K models released in 1952. Meanwhile, the age of the foot clutch and the hand gearshift had passed, thanks in part to English competitors, and the Harley-Davidsons were now built using a hand clutch and a foot gear-change control.

The "KH" could achieve 56 horsepower, and its engine and four-speed gearbox, which was a bit weak, were encased in a solid block. The gear shift was located under the right foot and the clutch consisted of a lever fitted near the left handlebar grip. Despite the qualities of this motorcycle, the American customers, and the young in particular, could not resist the temptation of owning a British bike. In 1953, the direct American competitor, Indian, shut down: constant change in its management had led the company to bankruptcy. Harley-Davidson was now alone to face the strong foreign competition.

Model K left many customers disappointed, but the company was not daunted by one failure. The groundwork to create a light motorcycle had been laid; but time was short and a lot of work was necessary to make it more reliable. In 1954, Harley-Davidson released the Series KH Sports Twin, whose displacement of 55 cubic inches is equal to today's "883 cc." In the following year, the KHK Super Sport Solo was introduced. The external appearance of the "KHK" was similar to that of the "K," apart from the caster angle, which had been reduced to provide better stability at high speeds but it came fitted with a special kit which increased its performance. The bike could reach 100 mph with 38 horsepower.

Production lasted for just two years.

***105 top** The 1952 Model K, a modern version of the old W, was fitted with a foot gear selector and a handlebar clutch. Its gearbox was integrated into the crankcase.*

***105 center** The XL Sportster, introduced in 1957, anticipated future generations of this consistently successful model.*

***105 bottom** The small S Lightweight, introduced in 1948, was sold for 13 years, during which it was only slightly modified.*

In 1956, a combined total of 1,253 “KH”s and “KHK”s were made and, in the same year, besides the two above mentioned models, the company introduced the specialty models the “KR,” the “KRTT,” and the “KHRTT,” all of which had a displacement of 45 cubic inches and were sport bikes for cross-country races. In 1958, these models stopped appearing in the catalog because, although the company still produced a few of the Ks, the new XL far outstripped them in terms of sales. Interestingly, this new motorcycle resembled the KH in certain details, such as the engine base. In 1958, the “XLH Sportster” underwent some important modifications: compression went from 7.5:1 to 9:1 and the valves became larger. Some Harley-Davidson dealers on the West Coast realized that this new motorcycle had interesting sports possibilities; at their request, Harley-Davidson conceived a lighter competition model.

Over the years, the Sportster underwent several technical improvements and has remained one of the top models in the Harley-Davidson line. Certain parts were later made in aluminum; new paints allowed it to preserve its youth and to adapt to customer taste. The introduction of discs meant changing the braking system slightly, and the electrical system changed to 12 volts. In 1965, the "XL" became "XLA," but only for the 1,000 units built for the U.S. Army. Subsequently, the kick-starter was abandoned, an electric starter was provided, modifying the sump, and the magneto was replaced by a coil ignition. In 1970, one year after the partial buyout of Harley-Davidson by American Machine & Foundry, came the launch of a model exclusively for sport, the "XR 750," which generated a buzz on the circuits and which many enthusiasts still dream of acquiring.

By 1972, over 82,000 Sportsters had been made since the model's release in 1957. Well conceived and reliable, since its creation, this model has been adapted to suit many types of riders and their myriad requirements. Today, it is still so fashionable that the new Harley-Davidson customers opting for a Sportster as their first bike very rarely know that its basic conception has existed for more than fourty years. It is also the least expensive model in the range.

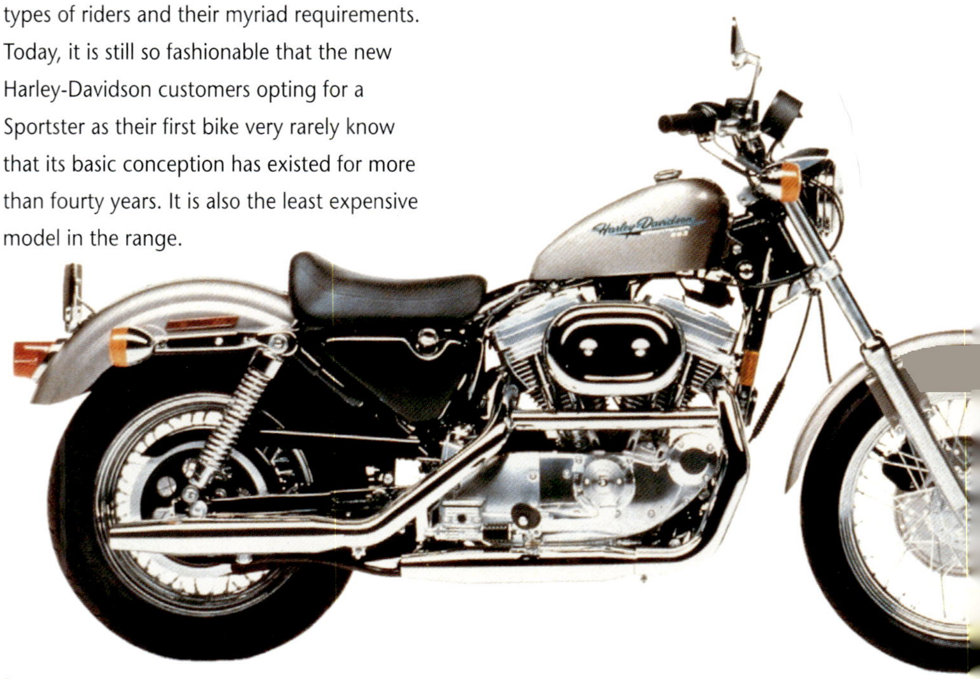

The "XLCR," a model in the Sportster line that was launched in 1977 with the name of "Café Racer," ended up being hunted by collectors because only 1,201 were made. Today, in exhibitions dedicated to collector's vehicles, this bike may fetch a price similar to that of a new "FLHTC."

In 1982, these light motorcycles underwent a great change, especially concerning the frame, which became even lighter and, in particular, more rigid. It also became capable of supporting a new tank with a 2 1/2 gallon capacity. In 1986, the Evolution engine appeared on the XLH 883 Sportster and XLH 1100. But in 1988, XLH 1100 was replaced by the 1200 XLH to provide more torque and power, and so to further differentiate it from the 883. At present, the Sportster 883 and 1200 are continuing their careers, albeit in slightly different versions, thus enriching the Harley-Davidson line.

106-107 The Sportsters, which appeared in 1957 with a 54 cubic-inch (883 cc) overhead -valve version, were the successors of the Flathead-type model K, although they were fitted with a crankcase including bottom engine and gearbox. Sportsters have been the company's longest-lasting and most consistently successful models. Through the years, their technology has improved and their range has increased. In 1986, a whole aluminum engine called the Evolution was unveiled, similar to the aluminum Big Twins sold from late 1983 on. Sportsters with Evolution engines were immediately available in 883 cc and 1100 cc versions.

1953 : Indian Disappears, Harley Remains

In 1958, the heavy Harley-Davidson V-Twins underwent some important changes. The Harley-Davidson Company was aware that it was important to keep in the catalog a style of bike that had helped build the brand's image over the years. Even today, some people still associate the Harley-Davidson brand with a white motorcycle, such as an "FLH," with wide handlebars and a rear wheel with either rigid fiberglass or leather bags at the sides, ridden by a stiff, proud policeman on a solo seat, dressed in spotless new clothes; in other words, an old Hollywood image. However, beginning in 1941, this "FL" model was actually subjected to many modifications, although its displacement remained 74 cubic inches (1200 cc).

First of all, as mentioned earlier, the Knucklehead engine gave way to the Panhead in 1948. In 1949, Hydra-Glide telescoping hydraulic forks became standard. The FL series was augmented by the addition of the FLH in 1955, and, in 1956, this series was composed of six motorcycles. Despite their similar appearance, a few small but key differences helped distinguish one from the other. Models "FL," "FLE," "FLEF," and "FLHF," whose total production amounted to 5,806 in 1956, accounted for almost 50% of that year's output from the factory in Milwaukee. In 1958, the Hydra-Glide was renamed the Duo-Glide (in turn anticipating the Electra Glide), and the new model was fitted with rear shock absorbers linking the frame to the swinging arm. This modification increased comfort and improved the heavy bike's ability to hold the road; but the rear shock absorbers had not definitively been adjusted. Their position was still inclined too far forward, reducing their efficiency considerably. However, because the seat was fitted on two springs and the large tire ensured a great deal of comfort already, the designers of this new "FL" did not make life too difficult for themselves by undertaking complicated studies concerning the fitting of the shock absorbers.

The Duo-Glide, equipped with this practical, if imperfect new feature and a hydraulic brake on the rear wheel, benefitted immediately, not least because the bikes, which were less stressed, had an increased road life.

In 1958, a small 125 cc model, the "Hummer," appeared in the catalog and stayed until 1960.

At that time, Harley-Davidson had been around for fifty-five years and had experienced a happy period before the Great War, a lively battle against its American competitors, the Great Depression, a second World War, and finally, the arrival of foreign competition – all events during which four men, joined by a very solid friendship and driven by the same passion, had been able to create, nurture, and build what had become the oldest motorcycle factory in the world. Then, as the years went by, the new generation followed and even-greater difficulties

108 The Panhead appeared in the Harley-Davidson catalog with this shape until 1965, proving its versatility. It was the first Harley-Davidson engine to be adapted to Chopper frames.

109 top and center The Duo-Glide replaced the famous Hydra-Glide in 1958. The main modification, from which the new name originated, was a rear frame equipped with dampers, giving a double shock-absorbing effect, one in the front from the telescoping fork and one at the back.

109 bottom The Enthusiast continued to boast Harley-Davidsons' merits, advertise accessories, and present new models.

***110 top** Police corps used Harley-Davidsons, equipped with all the accessories they needed: radios, windshields, warning lights, sirens, saddlebags, etc.*

***110 bottom** In 1958, accessories appeared such as saddle rails, which anticipated real seat backs mounted on sissy-bars.*

began to present themselves. Consumer needs evolved as motorcycle owners became more knowledgeable and demanded new and updated products. The Harley-Davidson Motor Company was lucky to maintain a core group of loyal supporters who kept selling and riding their motorcycles but even so, change was on the horizon. The company may have just celebrated its 50th anniversary, but twenty difficult years were ahead.

The November 1959 issue of Floyd Clymer's magazine *Cycle* announced the arrival of a new brand of motorcycle on the American market, but it didn't predict the impact that arrival would have on the motorcycle field of the future. It was no longer a question of the danger related to the competition coming from Europe, but to that from Japan and a brand called Honda, which was hardly known.

Penetration of the market was very modest in the beginning. Honda very wisely settled on American soil with a small 50 cc motorcycle. This small capacity motorcycle worried no one, and most people didn't even pay attention to it. At the time, Japanese material was not renowned for its reliability.

The Japanese strategy was actually quite well thought out: they invaded the American motorcycle market by offering products at extremely low prices. Although this did not bring a huge profit, it was a cost-effective means of establishing brand recognition. Then, once the brand was well established, they began exporting heavy motorcycles, which often even copied American design.

There are those who think that Harley-Davidson played an important role in the success of the Japanese motorcycle industry. In 1929, when the yen lost half its value, American imports, notably the Harley-Davidson, became much too expensive for Japanese pockets. In 1932, Harley-Davidson sold a manufacturing license to the Japanese company Sankyo (at the

111 top Along with other public authorities, many police departments used Harley-Davidson Servi-Cars until the 1970s. The production of these side-valve engine vehicles stopped in 1974.

time a pharmaceutical company, but a brand which can still be found on electronic equipment today) to manufacture a sort of Japanese Harley-Davidson, called "Rikuo" (King of the Road).

In the 1960s, when Japanese motorcycles began posing a serious threat to the last U.S. manufacturer, it was up to William Herbert Davidson and his associates to devise a strategy for preserving the future of the brand.

111 center and bottom Duo-Glides which have been renovated and fitted out with historically accurate accessories are often shown in competitions. Many owners do not hesitate to improve the Duo-Glides' looks even further by using newly chrome-plated parts.

Loss of Personality

Harley and the Italians

In the summer of 1960, the Harley-Davidson Motor Company started to negotiate with Aermacchi, an Italian company based in Varese. This step was aimed at finding an arrangement which would allow Harley-Davidson to distribute lightweight bikes that would compete effectively against the Japanese models.

Before World War I, the Aermacchi company, which had been based in Milan since 1912, manufactured planes in small quantities.

At the beginning of 1930, as orders came to build warplanes for Mussolini's airforce, the company grew. Then, during the Second World War, it was almost destroyed. After the war, it reestablished itself more modestly in Varese and began to manufacture light cars; but it continued to suffer financial problems. Aermacchi thought motorcycles might be a more profitable business, and began manufacturing lightweight models. In 1960, Harley-Davidson purchased half of Aermacchi's bike division, forming Aermacchi/Harley-Davidson.

In 1960, as a result of the agreement with Aermacchi, Harley-Davidson marketed a scooter, the "Topper," equipped with an automatic gearbox, a belt drive, and a 165 cc engine. It remained in the catalog through 1964, but was not a great commercial success.

During the first years of cooperation with Harley-Davidson, Aermacchi took part in important races with a 250 cc motorcycle designed specifically for the purpose. In this period, other Italian models were also put on the American market and, in 1962, Harley-Davidson launched the BT Pacer Lightweight, a 175 cc motorcycle. This model was reliable and had a quality in line with the reputation the factory in Milwaukee had been forging for sixty years, but the competitors were simply offering more appealing and aesthetically sophisticated products.

(Incidentally, all through these years, Harley-Davidson realized that fiberglass was gaining ground and that the use of this material had become nearly indispensable to chassis conception and to certain motorcycle accessories. In 1962, the company acquired 60% stock in the Tomahawk Boat Company in Tomahawk, Wisconsin, and reorganized it to make it suitable for the production of motorcycle elements. Harley-Davidson also used the opportunity to manufacture small electrical golf carts in this factory, but not for very long.)

1965 marked the end of the Duo-Glide, which was ousted by the Electra Glide and went to the museum to join the Hydra-Glide, the model which it in turn had once replaced.

The Electra Glide, whose name came from the fact that it had an electric starter motor, nevertheless preserved the kick-starter system. Its battery was 12 volts, rather than 6, and front and rear disc brakes replaced front and

***112 top left** William J. Harley, Walter C. Davidson, and William H. Davidson unveil the first machines coming from the Italian company, Aermacchi/Harley-Davidson.*

***112 bottom left** This advertisement capitalizes on the image of America in the '60s, promoting a small-capacity Harley-Davidson based on the 165 cc model released in 1953.*

***112 right** The Harley-Davidson Big Twins were used for official parades by police and the public authorities. The choice was only natural, since it was the last remaining American motorcycle.*

***113 top** The 1965 BT Pacer, with its two-stroke single-cylinder engine, appealed to younger people with leisure on their minds.*

***113 bottom** The first Shovelhead was put on the market in 1966. It consisted of a new engine top mounted on an engine bottom drawn from the Panhead, which is why it was later nicknamed "Pan Shovel."*

rear drum brakes. Until 1966, the motorcycle was equipped with a Panhead engine, which was then replaced by the new Shorelhead engine, which generated a great deal more power. In 1970, the Electra Glide was finally given its modern and definitive appearance when its engine sump acquired the shape with which we are acquainted today and when it abandoned its magneto in favor of an alternator. This bike was also later equipped whit a new carburetor – in short, it benefited from many improvements. As for the Sportster, on one of its versions, in 1967, it was fitted with an electric starter.

In the early 1960's, the third generation, which had already been involved for a few years, was asked to accept greater responsibility for various sectors of the company. A few years earlier, William H. Davidson, who had been president of the company since 1942, suggested that his son, William G. Davidson, take on responsibility for the design department. In 1963, William G. accepted this offer; it gave him the opportunity to use his skills to contribute to the factory that he still considered a part of the family. It is worth pointing out that Willie G. had graduated from an art and design school in Pasadena, California. Before working for Harley-Davidson, he had already demonstrated his talents at Ford, the same company that had seriously worried his grandfather and his uncles at the beginning of 1900. Having grown up among bolts, wheels, and V-Twin engines, and having taken part in endurance motorcycle races, he was aware of the needs of the company's customers. He accepted his father's offer and started working immediately. Although it seemed a good sign, Walter Davidson's son Gordon died in 1967, perhaps precipitating some of the company's imminent difficulties.

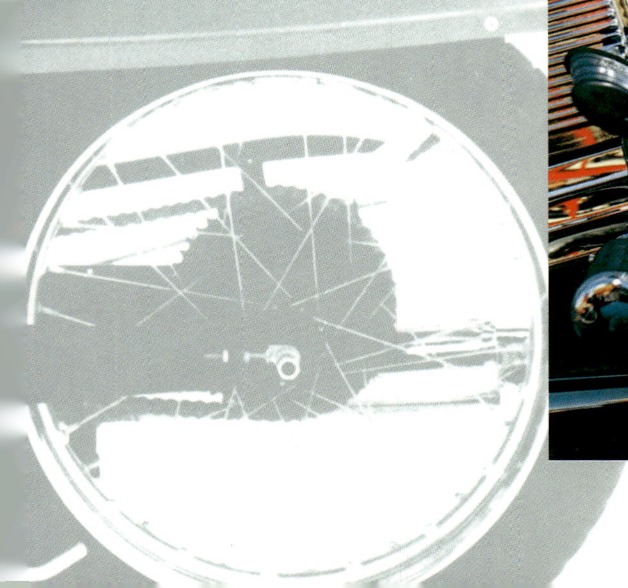

Harley Faces Great Difficulties

In 1967, Harley-Davidson became the target of Bangor Punta, a firm working in the railroad sector and specializing in the buyouts of companies threatened by bankruptcy. To prevent the brand from disappearing, President William H. Davidson accepted a proposal from American Machine & Foundry (AMF) to acquire Harley-Davidson for the amount of $21 million. AMF would retain control for the next eleven years. Rodney C. Gott, who acted on behalf of AMF and felt a particular affection for Harley-Davidson, became a member of the executive management for the new company group, called AMF/Harley-Davidson. For Harley-Davidson, it was the end of sixty-two years of private, family management.

The financial injection was indispensable. Harley-Davidson was equipped with old machinery and was incapable of producing a large number of motorcycles quickly and cost-effectively. Therefore it was unable to compete against the Japanese. In order to fend off the pressure from Japanese motorcycles, which were continually being improved, it became necessary to modernize production systems. However, it was imperative that a rapid growth in production go hand-in-hand with quality and reliability, two attributes which had shaped the Harley-Davidson reputation for decades. Furthermore, the company needed to set up a research department which would be able to quickly satisfy customers who were increasingly eager for new technological and stylistic improvements.

Three letters now appeared beside the Harley-Davidson logo on the tanks: AMF, and Harley-Davidsons began to be mass-produced. It was the only way to supply sufficient quantities to retailers and sufficient variety to customers. Thanks to this relationship with AMF and the change it entailed, once again Harley-Davidson production began to grow; in 1969, it reached 15,575 bikes, then 16,669 in 1970, 22,650 in 1971, 34,750 in 1972, 37,525 in 1973, 40,430 in 1974, and finally 34,255 in 1975.

But many faithful customers, especially in Europe, did not understand the change. Why was their favorite brand making bikes that were so indistinguishable from their competitors? What had happened to the unique Harley spirit? Rodney C. Gott soon understood that the production increase had been accompanied by a significant drop in quality and that Harley-Davidson could not compete against the Japanese on their own ground – that of mass-production. He inferred, from this analysis, that the only solution for the American brand was to

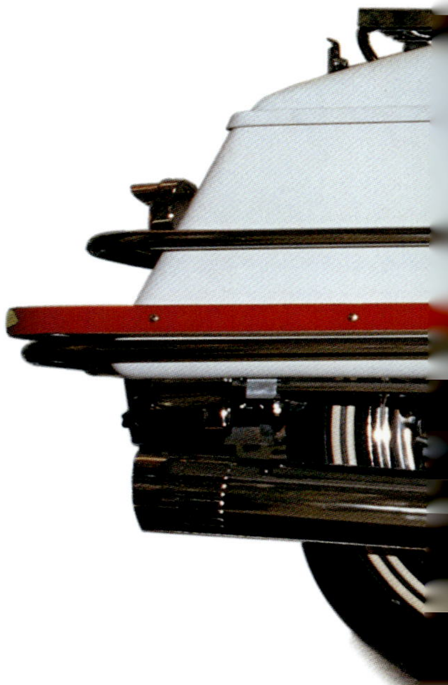

***114 bottom** The Sprint, a small four-stroke 250 cc motorcycle, was sold until 1969, when the capacity of the road model was increased to 350 cc. Made in cooperation with the Aermacchi/Harley-Davidson factory, this model would disappear from the line in 1975.*

***114 top** The MSR-100 appeared in 1971, under the name Baja 100. This small two-stroke 6 cubic-inch (100 cc) Harley-Davidson was not able to win the competition against the Japanese.*

***114 center** Harley-Davidson tried to penetrate the scooter market with the Topper. Introduced in 1960 and fitted with a two-stroke 165 cc engine, it was listed in the Harley-Davidson catalog until 1965.*

keep developing the trump cards which had built its success and its legend: solidity, reliability, and quality. During this period, the Japanese reinforced their penetration with models which were as big as the Harley-Davidson V-Twin. In 1975, Honda presented the "Gold Wing" in the United States, a massive motorcycle with a flat 1000 cc four-cylinder engine, whose bulbous shape was close to the legendary Harley-Davidson.

Rodney C. Gott believed that Harley-Davidson's problems were caused by the Harley-Davidson family management, but the events that followed and the succession of different presidents proved that the source of the company's trouble lay elsewhere.

William Herbert Davidson had held the office of president for twenty-nine years; in 1971, he left the post, although he continued to work within the company as chairman for two years. In 1973, when he retired, one of his sons, John A. Davidson, Willie G.'s brother, took up the

***114-115 and 115** Harley-Davidsons fitted with Shovelhead engines gradually stood out as deluxe touring motorcycles. They appealed to those who were eager to ride through America's wide open spaces on a motorcycle that provided maximum comfort. These riders did not hesitate to decorate their motorcycles with chrome plating, accessories, and specially designed radios.*

116 left In 1976 Harley-Davidson celebrated America's bicentennial by selling machines with "Liberty Edition" tank decor. Posters commemorated the event, presenting Harley-Davidsons as American machines linked to images of freedom and escapism.

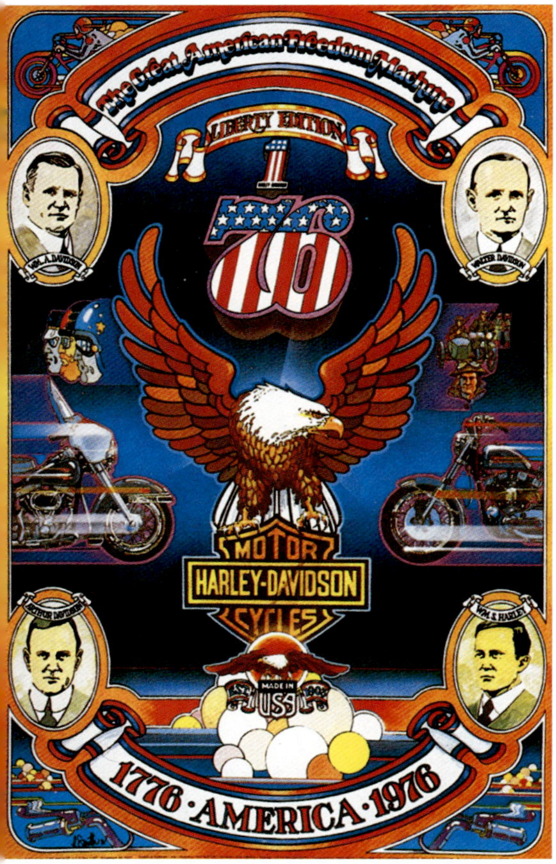

116 right The first Shovelhead engines appeared in 1966. They provided a ten percent power increase over the Panhead, while ensuring better lubrication and greater reliability.

117 Production of the SS 250 started in 1975. Equipped with two-stroke one-cylinder engines, these motorcycles were available in road and cross-country (SX) versions, with 175-and 250-cc V-Twin engines.

presidency. In the same year, the line was widely modified, both in terms of decorations, with new colors and logos, and in terms of frames and tanks.

But the factories in Milwaukee and on Capitol Drive were incapable of concurrently implementing these modifications and of increasing production, which amounted to more than 70,000 motorcycles in 1973. AMF owned a large factory producing military material in York, Pennsylvania. The management decided to move the manufacture of frames and assembly of the Harley-Davidson line in this factory, leaving the manufacture of engines and gearboxes in the Capitol Drive plant and in Milwaukee. The first motorcycle to come off the York line was a Sportster, one of the best-selling models of the Harley-Davidson range.

Unfortunately, this separated manufacturing process led to myriad problems. While production increased in York, the factory in Wisconsin could not keep the same pace. Because each factory was managed in isolation, according to its own imperatives, there were problems in communication and coordination, and, consequently, irregular production.

Although some customers remained loyal, AMF/Harley-Davidson lost a lot of money, because the finished motorcycles revealed an increasing number of defects. As a result, in an increasingly tense atmosphere, AMF started to change the members of its management team, which led, finally, to a three-month strike. It proved difficult for the new head, Ray Tritten, to find reasonable solutions. When Ray Tritten took over, there was no distribution plan, and stocks were badly managed; not to mention the complex production system and the divisions

between managers. Tritten was able to solve certain problems.

The only solution for Harley-Davidson was to improve the quality of its machines and make customers forget the simplicity of its mechanics as compared to the sophistication of Japanese bikes. (While it was true that Harley-Davidsons started easily, braked better and practically did not vibrate, the press dedicated only a very few articles to the American motorcycles, arguing that they lacked new features compared the continuous innovations of their foreign counterparts.)

Ray Tritten thought the best solution was to appoint a new head in Milwaukee, one capable of finding and introducing a new management policy likely to set this legendary company back on its track. After a search, L. Vaughn Beals, a disciplined and uncompromising man, was

appointed president in 1977. He was completely new to the world of motorcycles, but was able to understand the problems of a firm whose structure was so disorganized. The company needed a long-term policy rather than a day-to-day strategy; it needed to improve the existing motorcycles by making them more reliable; and to fight against the competitors with machines of equal performance. To this end new, technologically improved engines needed to be developed, without sacrificing the personality of the famous V-Twin.

Jeff Bleustein, executive president from 1998 to 2009, was chief mechanical engineer at the time. He was put in charge of setting up a research department for Harley-Davidson. Jeff Bleustein thought it was essential to keep the characteristic sound of the V-Twin. A Harley-Davidson can be recognized from afar by this noise, which identifies the brand even before the water-drop shaped logo on the tank is within sight.

In the early 1980s, the management set up a project called "Nova," aimed at improving the V-Twin and then creating a new range of bikes and engines to compete against the Japanese. Today, these two objectives would appear almost impossible to accomplish, because a Harley-Davidson has really become inseparable from the V two-cylinder engine. However, at that time, Vaughn Beals and Jeff Bleustein easily obtained the budgets needed for the research from AMF and asked Porsche, in Germany, to study a new engine. Ten million dollars were spent on the "Nova" project. It ended with the abandonment of the Porsche prototype, which was built around a V 4 engine block.

In 1978, the research offices at Harley-Davidson were given an imperative: propose a design for a new engine, differentiating it from the Shovelhead. So, its rocker arm cover could not resemble that of the Shovelhead, and obviously, its production cost had to be reduced in comparison to its predecessor. It took 250,000 miles of testing this new V-Twin to reach the desired reliability before it could be fitted on the Harley-Davidsons, beginning in 1984. And the V2 Evolution was born. These six long years would demand a tremendous amount of tenacity from Harley-Davidson, as well as the strength to embark on a new adventure.

Personality and Independence

The Arrival of a Magician

Harley-Davidson has many faces, and while older people may think of police motorcycles when they hear the name, for younger generations it often conjures up the image of a chopper, the machine ridden by the bikers of the '60s and '70s. The film *Easy Rider,* starring Peter Fonda and Dennis Hopper, showed precisely this kind of motorcycle, in this case a completely transformed Electra Glide.

In 1971, Willie G. Davidson had been responsible for design within the company for eight years. After observing this new way of conceiving the motorcycle and especially the Harley-Davidson, he took his pencil and sat at his drawing board. For the sake of reliability, he conceived a completely new but very economical model, which modified various elements of the models included in the range that year. The thin, light Sportster front and the heavy, stable Electra Glide back gave rise to a bike capable of satisfying different kinds of customers, but above all of injecting new life into the ranks of Harley-Davidsons. The "Super Glide" did not particularly surprise the customers. The tank shape was still that of a water drop, the fork was still thin, the head-lamp simple, the handlebar like a cow horn, the twin-seat located on a streamlined and shaped mud-guard; its colors were borrowed from the American flag, consisting of a white background with red and blue geometrical and symmetrical patterns. These modifications merely modernized the appearance of the Harley-Davidson, which remained practically the same in terms of its engine and frame. Its engine was a 1200 Shovelhead, so this bike could be considered the mixture of two motorcycles.

In 1972, the Sportster engine reached 1000 cc, thanks to a larger engine bore. In 1974, the Super Glide was given an electric starter as an option, and the range was completed by the "FXE," a designation indicating the motorcycle fitted with this new accessory. Nevertheless, the standard FX model continued to be produced until 1978, a year in which 1,774 items were manufactured; up to 8,134 "FXE"s, were produced in the same year. All in all, the factory produced 79,700 Super Glides between 1971 and 1978.

The Super Glide lead the way for a whole series of models derived from that first creation of the design department head. So, in 1977, it was followed by the "Low Rider," then the "Fat Boy," and finally, in 1980, the "Wide Glide," which became one of the company's best-selling motorcycles.

Willie G. quickly understood what the Harley fans expected; they were not necessarily looking for fast motorcycles, but rather for machines with character and power, which would be able to maintain a certain style and personality over the years.

118 top ***In 1974, Willie G. Davidson (left) and William H. Davidson presented an entirely renewed Racer to Tony Hulman, who would show it in the museum in Indianapolis. Joe Petrali had been national champion six times on a motorcycle like this.***

118 bottom ***In 1971, AMF/Harley-Davidson presented this Super Glide with its original design, particularly daring in those days. This motorcycle marked the beginning of what was then called "factory custom-ization," with Harley-Davidson offering machines with special lines that distinguished them from mass-produced motorcycles.***

In 1984, after the Wide Glide, the Softail was put on the market. Since it incorporated the characteristics of the Super Glide, the Low Rider and the Electra Glide, it initially gave the impression that Harley-Davidson was releasing a 1958 model for nostalgia buffs, but this was not the case; and although the frame seemed rigid, it disguised two shock absorbers which ensured a comfortable ride, one of the reasons today there is a complete Softail line built around this frame.

During the '70s, while Willie G. Davidson put together and designed some adjustments for the new models, AMF/Harley-Davidson experienced an impressive increase in production. But the Japanese still produced and redesigned more rapidly, thus obliging the dealers to drop the prices of the motorcycles left in stock. What is more, the American-Italian association found itself handicapped by the light motorcycles manufactured by Aermacchi, which were not able to stand up to the Japanese competition either.

In 1978, Rodney C. Gott retired from his post as AMF head and Tom York replaced him. Unfortunately, York did not understand the special relationship Harley-Davidson had with its customers.

As a result, York concluded both market share and profitability could be recovered if AMF built new models whose qualities and style corresponded to the Japanese fashion. Obviously, these proposals caused great concern amongst those who knew and loved the company.

Supporters of the company, who realized that the company's main strength lay in its personality, rallied togethet to put an end to the AMF and Harley-Davidson cooperation, so as to prevent the extinction of Harley-Davidson at the hands of Tom York. In 1980, Vaughn Beals tried to make the head of AMF understand that the sale of Harley-Davidson represented the best chance for its survival.

In 1983, the Nova project was finally abandoned. It was a huge financial loss, but where would it have led with such instructions? Surely to super-powerful, sophisticated but unoriginal motorcycles; to simply entering into direct competition with the Japanese manufacturers; to trying to be little more than the company that releases the fastest model and the fanciest fairing.

***119 top** The entirely black 1977 Café Racer was another customization of a Sportster. Unluckily for Harley-Davidson, this conception of the sports motorcycle was too advanced for those days.*

***119 bottom** With this 1977 FXS, called the Low Rider, factory customization continued. The Low Rider met Harleyists' expectations more closely than the Café Racer of the same year. Thanks to its success, more models were derived from this motorcycle.*

Independence and Rescue

Vaughn Beals, the son of an accountant from Boston and the person who would lead Harley-Davidson's return to independence, had studied aeronautical engineering at the Massachusetts Institute of Technology. He had worked for various companies before becoming a research and development engineer at and then head of an important aeronautical technology factory. Thanks to his diplomacy, once at Harley-Davidson, he managed to reinforce the team spirit and suggested that the other members of the executive management take on more responsibility. He motivated and helped the dealers by granting them some discounts.

His objective was to consolidate the customers who were still loyal to the brand and to lay the groundwork for a new start for the firm. He launched the 5-speed gearbox and various other improvements, but unfortunately, at the end of the '70s, financial problems meant many projects had to be abandoned. Eventually, Vaughn Beals left his post, but remained within the AMF group.

Then, on February 26, 1981, Beals gathered twelve other Harley-Davidson executives disappointed by the association with AMF to suggest that they should buy the company together. He knew that the asking would not be very high, because Harley-Davidson was considered beyond recovery. Vaughn Beals started discussions with Citicorp, the industrial credit company; these proved very difficult, as were the discussions with AMF.

By this time, Harley-Davidson had had six presidents in eight years. When Vaughn Beals launched the redemption operation, Willie G.'s brother, John A. Davidson, who became the head of the golf cart department, retired from the management team, thus leaving Willie G. as the last remaining member of the big family that had started the eighty-year-old enterprise.

Nonetheless, the difficult negotiations paid off, and on June 14, 1981, the contract of sale of Harley-Davidson was signed by the AMF president, Ray Tritten. At that time, the thirteen board members were as follows: John Hamilton, Jeffrey Bleustein, Kurt Woerpel, Chris Sartalis, Willie G. Davidson, James Paterson, Timothy Hoelter, David Lickerman, Peter Profumo, David Caruso, Ralph Swenson, Charles Thompson, and Vaughn Beals. Together they got on the road to make the long trip from York to Milwaukee.

***120 top** With the signing of this document, the Harley-Davidson Motor Company recovered its independence, in June 1981.*

***120 bottom left** The new owners and full-time employees of the Harley-Davidson Motor Company: (standing from left to right) John Hamilton, Dr. Jeffrey Bleustein, Kurt Woerpel, Chris Sartalis, William "Willie G." Davidson, and (sitting from left to right) James Paterson, Timothy Hoelter, David Lickerman, Peter Profumo, David Caruso, Ralph Swenson, Charles Thompson, and Vaughn Beals.*

***120 bottom right** The first motorcycle produced by the independent Harley-Davidson Company was a Heritage Edition, decorated with a gold medallion on its oil tank cap.*

***121 left** The Sportster XR-1000 evoked the racing XR-750 models produced for competition. This motorcycle, which became available as a limited series in 1983, contributed to the recovery of Harley-Davidson.*

***121 right** The 1981 Heritage Edition, one of the first Harley-Davidsons other than the Electra Glide to come with a 1340-cc capacity. Although it was fitted with a modern Shovelhead engine, it developed a totally retro look, anticipating later Harley-Davidsons equipped with Evolution engines entirely made in aluminum.*

Harley-Davidson had returned, with difficulty, to a unique identity.

After this historical signature, some people criticized AMF very harshly for not taking better care of the classic company which had been in their stewardship. Others reacted more reasonably, taking into account the efforts AMF had made: if AMF had not invested money to develop the new Evolution V-Twin, or simply to build of the Super Glide and other newly conceived models, Harley-Davidson would have disappeared long ago.

At the moment the company recovered its autonomy, the motorcycle market in the United States was not at all favorable. Production fell again, to the consternation of its new executives. Additionally, certain Japanese motorcycles that had the same aesthetics, if not the same spirit of the heavy V-Twin arrived on American soil. They were sold for half the price charged by Harley-Davidson Motor Company.

At the end of 1981, Harley-Davidson found itself in the worst situation the company had ever known. Since certain parts became very difficult to find, most of the Harley-Davidson dealers opted for the sale of Japanese motorcycles and only a few remained loyal to the brand, perhaps as a result of some premonition. By waiting for better days, they showed their confidence in the thirteen executive managers.

In Japan, inventory was updated several times a year, allowing a thorough inspection of the parts in terms of quality and cost price. The ideal thing was to have the right parts at the right moment. It was also a matter of further involving the worker, of making him feel responsible, so that he watched over the execution of each assembly step and avoided becoming a robot. He had to check his production pace in order to remain in step with the rest of the manufacturing process. Each operation had its name.

1. The "EI" (Employee Involvement) was the complete participation of each employee in problem solution and quality control.

2. The "JIT" (Just In Time Inventory) was the reduction of expenses and quantities.

3. The "SOC" (Statistical Operator Control) was each employee checking his own work.

Harley-Davidson learned from the Japanese and implemented these three organizational and regulatory systems. Traditionalism and conservatism also had to be put aside in order to make way for a new method.

***122-123 top** The classic image of the Harley-Davidson – fitted with the Panhead engine, linked to one- or two-up touring – was adapted to the fashion of the time in advertisements for new Harley-Davidsons in the 1980s and '90s. Advertisers used the same theme again for the Electra Glides.*

The suppliers themselves had to accept these new working formulas and opt for a pace that was identical to Harley-Davidson's, so as to be able to deliver on time and with the required quality.

After this reorganization, only the necessary quotas were manufactured each day and immediately assembled on each bike. The Harley-Davidson engineers responsible for the assembly, namely, those in the York, Pennsylvania factory, met regularly with the suppliers to discuss problems that had arisen during assembly or to define the qualities of an accessory worth following up. Thanks to the "Statistical Operator Control," the assemblers and the manufacturers could spot the specific problems at the beginning of production and stop the process before any motorcycles were rejected by the commercial service or by the customers. There were very few employees in the Harley-Davidson factories who had never ridden a Harley-Davidson. Many of them owned one, even several, and when working on this legendary line, they did their best to manufacture a flawless product. They managed this by unifying "EI," "JIT," and "SOC," the working methods inspired by their competitors – the same competitors who had nearly managed to eradicate the Harley-Davidson brand from the world of motorcycle production.

But Harley-Davidson still had to become competitive in terms of price. This was not simple, because the Japanese were already in a strong position and had nothing to prove. Thus, it was difficult for the company from Milwaukee to appeal to foreigners and especially to reconquer its domestic market, which was the key to ensuring the company's recovery.

In September of 1982, responding to the request of the managers at Harley-Davidson, the International Trade Commission took action for the first time to help the company. The I.T.C. submitted an official request to President Ronald Reagan aimed at increasing tariffs on imported Japanese heavy motorcycles (more than 700 cc) over a duration of five years (until 1988). The tariff amounted to 45% in 1983, 35% in 1984, then dropped to 10% in 1987, a year by which there were no more than 10,000 foreign heavy motorcycles on the American market. Since Harley-Davidson had regained its strength and the production pace of its best days, Vaughn Beals asked the government to lift the measure and to give free rein to the competition.

In 1983, at the beginning of this relaunching, only part of the factory was being used for the manufacture of motorcycles; Harley-Davidson seized the opportunity and began producing metal racks for carrying Air Force bombs. The contract with the military was actually the company's most profitable and, during Ronald Reagan's presidency, provided more than $20 million per year. In effect, 40% of the company profits came from a sector which accounted for 20% of production.

In 1984, the new V-Twin, "V2 Evolution," was introduced as part of the 1340 cc range. The Shovelhead, always a source of trouble, left room for a V-Twin capable of inspiring confidence in the Milwaukee motorcycle once again.

Some years earlier, Harley-Davidson had begun manufacturing special limited series of motorcycles to mark important American events, which was an efficient way to boost sales. The Milwaukee managers would even suggest to the retailers specific ways to promote the special editions. In 1986 came the Liberty Edition, celebrating the 100th anniversary of the Statue of Liberty, and on July 18, Vaughn Beals burst into the New York Stock Exchange driving a Limited Edition "Liberty" Super Glide to announce the Harley-Davidson share issue.

In 1986, Harley-Davidson bought the Holiday Rambler Corporation, one of the biggest manufacturers of leisure and commercial vehicles in America, which specialized in the manufacture of campers. The union lasted until the beginning of 1996, when Harley-Davidson sold back this department. The goal: to concentrate entirely on motorcycles, obtaining enough investment to increase production beyond 200,000 motorcycles per annum by the year 2003.

122 center The Harley-Davidson Heritage Softail first appeared as a limited series in 1985, then it became part of the line in 1986. Its components were based on modern technology but its looks were openly inspired by the old Harley-Davidsons.

122 bottom The Low Rider was released in a limited series: it was fitted with a Shovelhead engine, painted black, equipped with belt primary and secondary transmissions, and called the Sturgis model, after the fortieth anniversary of the Sturgis meeting.

123 bottom Once the Electra Glide adopted the Evolution aluminum engine, it finally achieved the success it deserved as the ultimate grand touring machine, rapidly conquering America as "King of the Highway."

Four Historical Factories, One Myth

***124 top and center** The Harley-Davidson factory, manufacturing engines and gearboxes, is located in Wauwatosa, a suburb of Milwaukee, Wisconsin. Motorcycles are assembled at another factory, in York, Pennsylvania.*

Harley-Davidson is now the leading heavy-weight motorcycle manufacturer in the world, but recovery came after disaster. When analyzing the difficulties that it encountered in the late 1970s and early 1980s, the company tried to accurately assess the technical measures that helped it regain success. It was aware that problems did not arise from personnel so much as its construction techniques and manufacturing system. In light of that understanding, as mentioned earlier, the company reorganized, following three strategies similar to those used by Japanese manufacturers: employee involvement, just-in-time manufacturing, and statistical operator control.

In 1978, even before the buyout, Harley-Davidson had become one of the first companies to start a quality circle program, encouraging employees to express their ideas, to solve problems, and to increase the quality and effectiveness of their own work. The company knew that only through its employees could it develop maximum efficiency. To begin with, each employee had to be recognized and respected. Then a team structure, organized around the company's goals, had to be established. Finally, a suitable incentive for achieving success had to be created.

That incentive involved the "just-in-time manufacturing" techniques, which Harley-Davidson labeled "MAN" (Materials As Needed). Before MAN, the company's

***124 bottom** The oldest Harley-Davidson factory, and the one that best symbolizes the brand, is still located on Juneau Avenue in Milwaukee. It now houses the company's administrative offices.*

techniques could lead to long delays, a big stock, and a rigid production system. MAN was a strategy by which the company could evolve.

First, Harley-Davidson developed a system to help it better predict customer demand. The company then reorganized its manufacturing system to group activities by category and reduce assembly times. The new system improved parts management and increased quality by allowing operators to follow a motorcycle through the whole manufacturing process, start to finish. It also permitted the rapid correction of errors and reduced the physical space needed for assembly.

The next step in reorganization was for operators to monitor parts constantly, identifying those needed and assuring more rapid restocking. With better control of parts acquisition, response rate increased and paperwork decreased.

These MAN policies could only be adopted with the full commitment and heightened awareness among all workers of everybody's responsibilities, at all levels of production. Through MAN, the company set up a real partnership with its suppliers: through the new "partners for profit" program they were given even more incentive to answer the needs of the company. This program helped encourage suppliers to maximize quality and reduce manufacturing costs, just as Harley-Davidson itself was. During this process, Harley-Davidson reduced the number of its suppliers by half. Those who remained were promised a long and fruitful partnership.

Statistical operator control (SOC) also proved its effectiveness day after day, by constantly improving production and reducing manufacturing costs. Thanks to this method, the company could rapidly correct emerging mistakes and constantly improve quality. Statistics became tools that operators could use to solve their most urgent problems. Harley-Davidson imagined that the program would only last three years, but it has actually become permanent, making it more possible for the company to respond to emerging economic and technical imperatives.

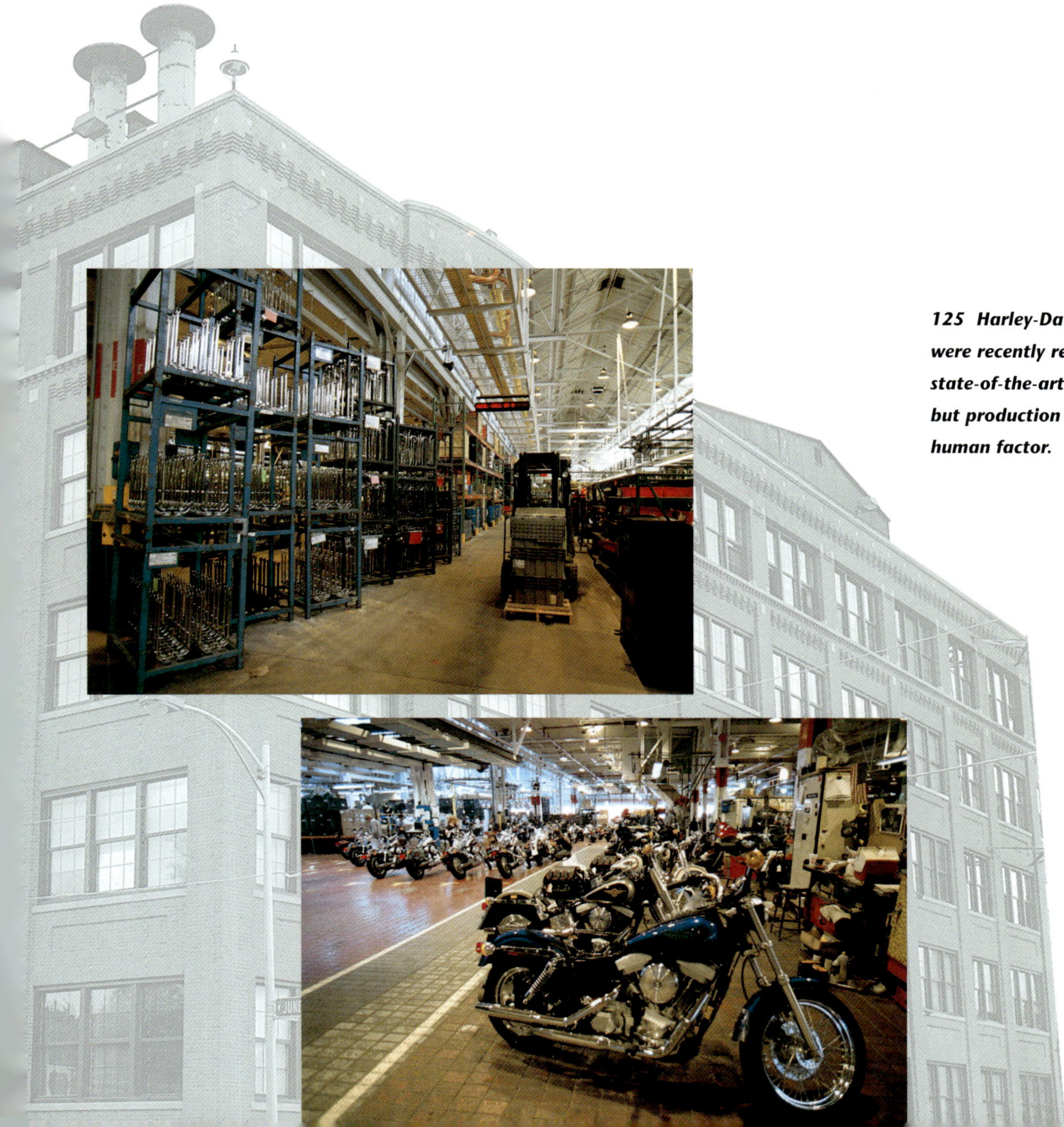

125 Harley-Davidson's assembly lines were recently reorganized, incorporating state-of-the-art manufacturing technology, but production still depends upon the human factor.

The Factory in Wauwatosa

Any neophyte Harleyist imagines that all Harley-Davidsons are built and assembled in the famous factory in Milwaukee, on the corner of Juneau Avenue and 37th Street. That was where the Harley-Davidson story began, but it is not the only location where the great motorcycles are now built. Nowadays, the midtown-Milwaukee building houses Harley-Davidson administration. Most actual manufacturing takes place elsewhere.

Since 1969, Tomahawk, Wisconsin, 150 miles from Milwaukee, has been the manufacturing center for all the Harley-Davidson glass and fiberglass elements, including fairings, covers, saddlebags, and sidecars. Harley-Davidson bought the factory, once used to build boats, to benefit from experienced personnel and to avoid relying on an outside company for parts.

One challenge faced at the Tomahawk manufacturing plant was to match the colors of metal and fiberglass parts. Early on, such matches were impossible, which explains why, in the 1970s, the Electra Glide's fork head and bags were black and white. Today, this problem no longer exists, and fiberglass bags and other elements are almost exactly the same shade as mudguards and tanks.

As for true manufacturing, only the painting of engine parts takes place on Juneau Avenue. But foundries and assembly teams for the 1340 cc, 1200 cc, and 883 cc engines are located just a few miles away, in the Milwaukee suburbs on Capitol Drive. Close to a thousand employees work at this plant, manufacturing engine parts by hand. Strict planning avoids overproduction. Over the last few years, Harley-Davidson has also invested millions of dollars in automated manufacturing machinery, designed to produce reliable and high-quality parts. In the first area of the factory, engine parts are cast. Nearby, in the next area, men and women assemble those parts into engines.

Those engines will be fitted on motorcycle frames manufactured hundreds of miles away, in another factory in York, Pennsylvania. Engine assembly lines include quality control testing as well as manufacturing, guaranteeing

the reliability of engines as well as transmissions. On the V-Twin engine, for example, instrumented inspections are carried out intensively. The engines run for thirty minutes, then are dismantled and thoroughly checked. Certain tests are also conducted on V-Twins fixed to frames and made to react with transmissions, so as to simulate road behavior. Before being sent to the York factory, most of the engines get a coat of a special high-temperature-resistant paint. Chrome-plating and polishing take three-quarters of an hour; a finish layer of nickel and, sometimes, potassium bichromate is applied. Some parts – especially those with precise dimensions and high performance demands – are sent to York to be finished and tested.

In 1982, Harley-Davidson began welding frames, then assembling them in one operation, which ensured better-quality engine support and overall performance. Now these assembled elements are sent to 1425

126-127 The Wauwatosa factory, located on Capitol Drive, manufactures motorcycles' major elements, such as engines and gearboxes. Trucks deliver components daily, and the products they offload undergo daily quality control scrutiny. Similarly, at numerous stations throughout the manufacturing process, quality control experts examine parts, to preserve maximum reliability while increasing productivity.

Eden Road in York, where Harley-Davidsons come to life. The company first moved into this 400,000-square-foot facility in 1973, using it to assemble motorcycles and golf carts, manufactured during the time of the AMF association. Today the golf cart operation has been completely abandoned.

Harley-Davidson manufactures motorcycles according to market demand. In 1989, Harley-Davidson produced 55,000 motorcycles; of them, 8,100 went to Europe and 40,300 stayed in the American market. Production keeps growing. Volume of production in 1996 increased over that of 1995 by 12 percent, while 1995 volume had grown by 8.6 percent over 1994. In other words, production went from 490 machines per day in 1995 to 510 per day over the fourth quarter in 1996.

Most suppliers are based within one hundred miles of the Capitol Drive plant in Milwaukee. Since the early 1990s, 96 percent of the parts making up Harley-Davidson engines have been manufactured in the United States. The Milwaukee factory, like the one in York, operates on principles learned from Japanese car manufacturers. The Milwaukee factory tries to produce the exact number of engines needed in York to fill orders from dealers and other sales forces. Each working unit has exactly the material required by the assembly chain to fill this demand, neither more nor less. Working plans are drawn up daily and, for example, once a shelf of one hundred engine parts is empty, that shelf returns to its starting point, indicating that there are now one hundred engines on the line. Another shelf takes its place and an order goes to the supplier – no paperwork or management discussion required.

The company has been operating like this for years. It is a strategy that relies on workers, established after the recovery of the early '80s, as management acknowledged its past mistakes. Now all workers have their own responsibilities and try to do as well as possible, since everybody is interested in results. Each line operator is responsible for the part he is assembling; he must not let a faulty part go to the following station. Everybody accepts and appreciates the responsibility granted by this "quality circle" approach. Thanks to this principle, operation costs have been reduced, space is being used more efficiently, and maintenance has been reduced to a minimum. Moreover, assembly lines have greater flexibility, permitting, for example, special or anniversary models to be assembled without changing the factory and line operation.

Aware that motivated operators build company strength, Harley-Davidson has encouraged teams on each assembly line. These teams can arrange the working space and the tools or equipment as they like, in order to obtain the best possible comfort and productivity. They plan their work by themselves, modernize and improve operations, and take charge of the tools and material they need. They work as individual units on the assembly line. Just as small

128-129 Harley-Davidson's engine manufacturing and assembly lines still depend on human operations, despite a big emphasis on automation. Company heads do not want to replace employees with machines, except in cases where machines can do dangerous jobs. Then, they believe, mechanization increases productivity without disrupting the team approach.

businesses, they control the supply of materials and work with motivation to obtain coherent and continuous production. Thanks to this system, Harley-Davidson production and quality have increased.

At all levels of the organization, everybody tries to produce a product of the highest possible quality and during meetings, new ideas for improving or upgrading the product or the factory always come up. While many companies units are moving towards complete robotization, this is not the path chosen by Harley-Davidson. Although it uses some robotics, it considers the need of human supervision and control essential. Harley-Davidson is a living machine with its own philosophy. Robots are primarily used to carry out dangerous tasks, but the human factor remains vital in the conception of the company and its machines. Maybe this is the key to Harley-Davidson's unique spirit. These products do not resemble any other motorcycles, and every new Harley-Davidson rider benefits from the quality.

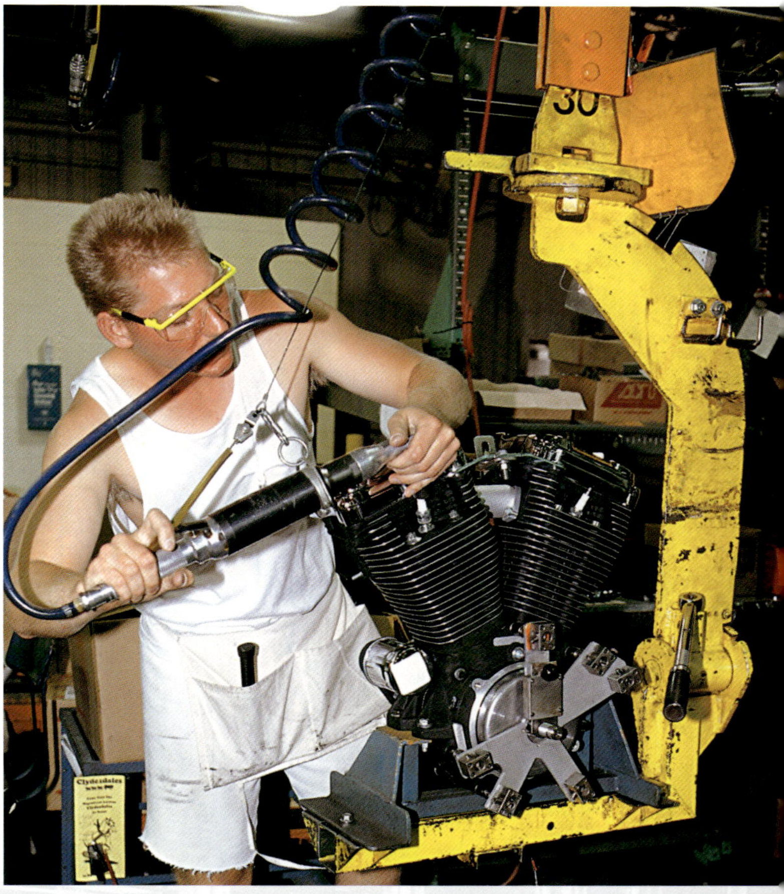

The Factory in York

The assembly factory moved to its new location in York, Pennsylvania, in 1974. The first thing a visitor sees is the warehouse where engines and transmissions arrive from Capitol Drive and wait their turn to be mounted on a frame, which determines the model. Boots, side bags, fork heads, and elements made out of fiberglass coming from Tomahawk also arrive at this warehouse.

Considering all the models under construction, several different stations are needed to assemble the hundreds of motorcycles completed here every day. As in Wauwatosa, parts arrive here as they are needed, so a large permanent stock has been eliminated. Certain elements made abroad as late as the end of the 1980s are now made in the United States, but a few elements, such as forks, carburetors, and electric circuits, still come from Japan.

The new system of just-in-time stock management has allowed a savings of nearly $40 million, not to mention the improvement of working conditions at York and the other factories. Each worker wears compulsory safety glasses as he works at one of the many machines. The biggest presses, hydraulically powered, are located at the beginning of the line; the most powerful have a force of 60,000 tons. These huge presses transform thin iron plates into mudguards and oil or fuel tanks, while others cut away excess metal from around the parts. Presses also fold the metal handlebars, leg guards, and bag guards to create the distinct models.

Wheeled shelves fill up with these recognizable elements. Then they are either chrome-plated or polished before painting. Next they advance to the station where the frame – the only Harley-Davidson part that passes through the entire York assembly line – is assembled. All the oil and fuel tanks are welded, polished, and carefully examined before reaching the dust-free painting booths.

The Harley-Davidson factory has invested a

***130 top** Since 1974, all Harley-Davidsons have been assembled in York, Pennsylvania. From 1998 on, however, Big Twins come out of the York factory and Sportsters are assembled in Kansas City.*

fortune in its ultramodern painting unit. At first sight, it looks like a completely aseptic space laboratory. Whether visitor or employee, everybody must wear a white overall, shoe-covers, and a cap to protect the hair, not to mention transparent safety goggles and face masks.

To visit this famous Harley-Davidson painting lab, one must walk through a room with a sticky floor, which retains all the impurities attached to the soles of one's shoes. Then one must walk through a flow of hot air from several ventilation ducts, so that all possible dust is removed. Then one enters the futuristic laboratory, where everything is automated and human intervention is limited only to the supervision of machinery and to the management of computer programs that run the equipment.

In short, mechanical operations accomplish the following procedures: the metal parts

are first accurately cleaned, through degreasing, washing, rinsing, and finally drying. Robotic arms, which look like spiders, keep parts in the right positions. Each of them can adapt to the special shape of any Harley-Davidson part in need of paint – tanks, mudguards, covers, etc.

After these preparatory stages, paintwork is also carried out by robots inside sealed booths. This factory is flexible enough to allow paint colors to be changed very rapidly.

***130 bottom and 131** On the assembly line, frames are first fixed on supports. Afterwards, they follow a track to various stations, receiving all the elements necessary to complete a Harley-Davidson by the end of the line. Machines then test the engine, gearbox, transmission, brakes, electrical system, etc., before the bike is packed for shipment.*

132 Few operations in Harley-Davidson manufacturing are carried out exclusively by machines. They support human work rather than replacing it, making workers available for other tasks, thus increasing factory productivity.

At the end, there is an inspection unit, where all the colors that have been applied are checked. Fresh, filtered, and moisturized air continuously flows through the painting unit for ventilation.

The maximum daily capacity of this unit is 400 motorcycles of one color or 270 if colors or models change mid-production. This factory also includes the laboratories that produce the paints, as well as quality-control labs that test the finished products. Each paint layer is baked in a furnace at a temperature of 300° for about twenty minutes before being covered with another finishing coat.

The body of certain models is decorated with colored borders.

In the penultimate operation, a sticker with the Harley-Davidson logo is applied on the fuel tank, and the whole motorcycle is painted with high-temperature varnish. In the final assembly line station, a large number of supports ready to be loaded with motorcycles moves slowly along a metal conveyor. A Dyna Glide, a Softail, or an Electra Glide frame, accompanied by an identification number, is solidly fixed to each support at the beginning of the chain.

Moved to the new Kansas City factory at the end of 1997, Sportsters have their own assembly line. Although Sportster

manufacture is different from that of the Big Twin, the operations are similar. Once the secondary transmission has been fixed, the V-Twin and its gearbox are fitted on the frame. By using jigs, the assemblers can tighten the bolts without risking damage to the engine. Several stations are needed to finally complete the assembly of the engine and the rear wheel.

A day of assembly at York is not dedicated to a single model. The Ultra Electra Glides follow the Softails, sometimes preceded by a few Dynas, all along the line. Some workers specialize in the assembly of forks and brakes. Others fix the radio and tape players onto the fork heads and do the wiring for tail and headlights, oil pressure and fuel gauges, the horn, and the starter.

At the end of the line, each motorcycle is examined. Test rooms simulate five miles of road conditions. Everything is inspected: brakes, shock absorbers, gears, horn, lights, and indicators. Motorcycles are placed on wheel supports, operated by the rear wheel. Then they are packed or remain in the hands of the engineers for a few hours, so that they can repair or replace a faulty part detected in the test room.

After quality testing, the Harley-Davidson motorcycles, enveloped in large, thick, transparent plastic bags, are placed on wooden boards. They are fastened down with

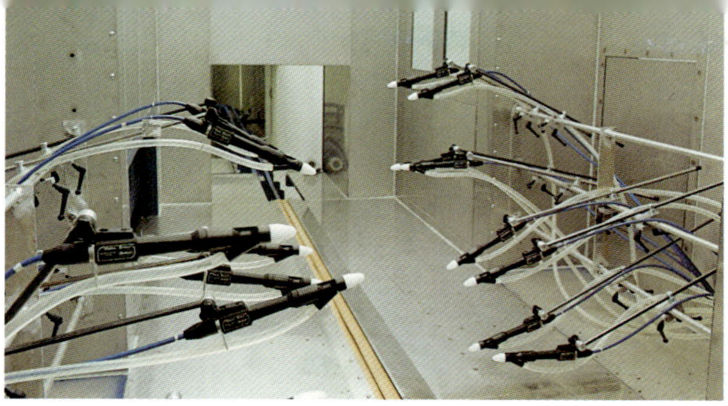

***133** The painting department of the Harley-Davidson factory in York is a top-quality unit, more like a laboratory than a common painting station and rarely accessible to visitors.*

belts, protected with small wooden arches, and packed in a huge cardboard box. Thus they leave the factory, bound for their final destinations in the U.S. or abroad.

In the United States, dealers can call on specialists who are trained to maintain and repair the various V-Twin models of the Harley-Davidson range. These mechanics follow three-month training courses in various schools; the main school, located in Phoenix and Orlando, is called the "Motorcycle Mechanics Institute." This school's training programs, sponsored by Harley-Davidson, are intensive courses, twelve weeks long, designed to provide a mechanic with the knowledge and experience needed to work on the engines, transmissions, clutches, shock absorbers, and electrical components of the Harley-Davidson. Courses are also taught in restoration, focusing on models dating from 1936 to 1969. Another school, the American Motorcycle Institute (AMI), founded in 1972 in Daytona Beach, Florida, trains mechanics to work on Harley-Davidsons, other brands of motorcycles, and boats as well.

In the middle of the 1990s, American dealers complained that they did not have enough Harley-Davidsons to sell. New arrangements enabled the company to increase production, reaching a goal of 200,000 motorcycles per year in 2003, the date of the company's 100th anniversary. Before the anniversary, many American customers have to wait several months for their Harley-Davidsons, depending on the models they have ordered – sometimes more than a year. Such a market situation goes far to explain the high prices currently paid for secondhand Harley-Davidsons, which are sometimes even more expensive than new motorcycles.

A Museum for a Great History: the Rodney C. Gott Museum

***134 top** The entrance to the Harley-Davidson Museum in York, same address as the motorcycle assembly plant.*

***134-135** The new 1997, FLSTS Heritage Springer stands out at the museum entrance, along with a sports model, the Buell, based on a Sportster engine.*

***135 bottom** The Harley-Davidson York Museum overflows with posters, newspaper clippings, leaflets, and some photographs showing the major executives of the Harley-Davidson Motor Company with celebrities and some not, as is the case here. From left to right: Ralph Swenson, Charles Thompson, Vaughn Beals, and William "Willie G." Davidson.*

Think Harley-Davidson, and you think the United States. These motorcycles are inextricably linked to the wide-open spaces – to deserts, to the Rockies, to Hollywood, to Texas, to the highway, and to dreams of escape, shared by generations of fans. Those who want to relive every twist and turn of the great Harley-Davidson adventure should visit the Harley-Davidson Museum of Milwaukee, which is opened since July 12 of 2008. Before, since June 8 of 1977, the Harley-Davidson Museum was the Rodney C. Gott Museum in York, Pennsylvania. These Museums exhibits tell the stories of the products and history of Harley-Davidson. If the RC Gott Museum was not so huge and with so many motorbikes like the new Milwaukee Museum, it was the first official and historic Harley-Davidson Museum. At the end of a large field, one can see some white-walled buildings, "York Division" in black letters on one of them. At the corner, a big gate opens to a simple entrance, then six steps climb to a double-windowed door. There you are, approaching the legend.

1906
UNIVERSAL TIRE, CO.

136 top, 136-137, and 137 top The first Harley-Davidson frame, with a curved front tube, was drawn directly from bicycle frames of the time. The original paintwork of the first models was black, red, and gold, but from 1906, gray also became available.

136 bottom A solid gold rendition of the first Harley-Davidson model, released in 1903.

1913
HARLEY-DAVIDSON

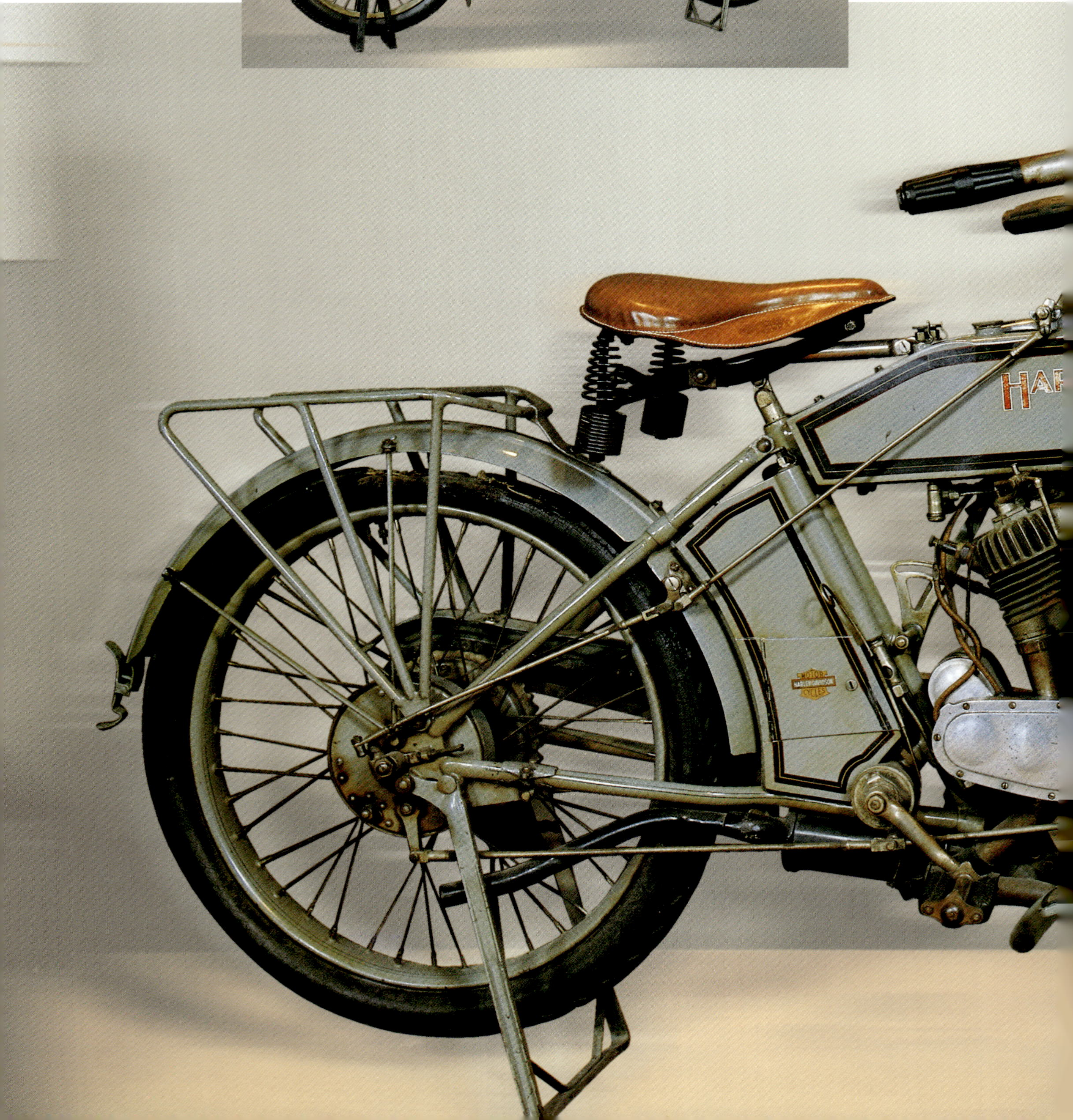
HAR

138-139 The V-Twin started its career with Harley-Davidson in 1909. The first available engine of this type had a 49.48 cubic-inch displacement. In 1912 it became available with 61 cubic inches (1000 cc) and a roller chain drive. The wheel size, ranging from 26 to 28 inches, could be changed as a function of owner weight. In 1914, footboards were introduced, along with a clutch and brake pedals.

140-141 The two-cylinder 61 cubic-inch 11 J model gained new pieces of equipment, such as a three-speed transmission with the lever fitted on the tank. This motorcycle had a sophisticated electrical lighting system with a two-bulb headlight, a taillight, a horn, and ignition. These new components were all optional, and most users preferred to keep the acetylene system, already fully tested. This model sold for $310.

HARLEY-
DAVIDSON
HARLEY-
DAVIDSON

HARLEY-
DAVIDSON
HARLEY-
DAVIDSON

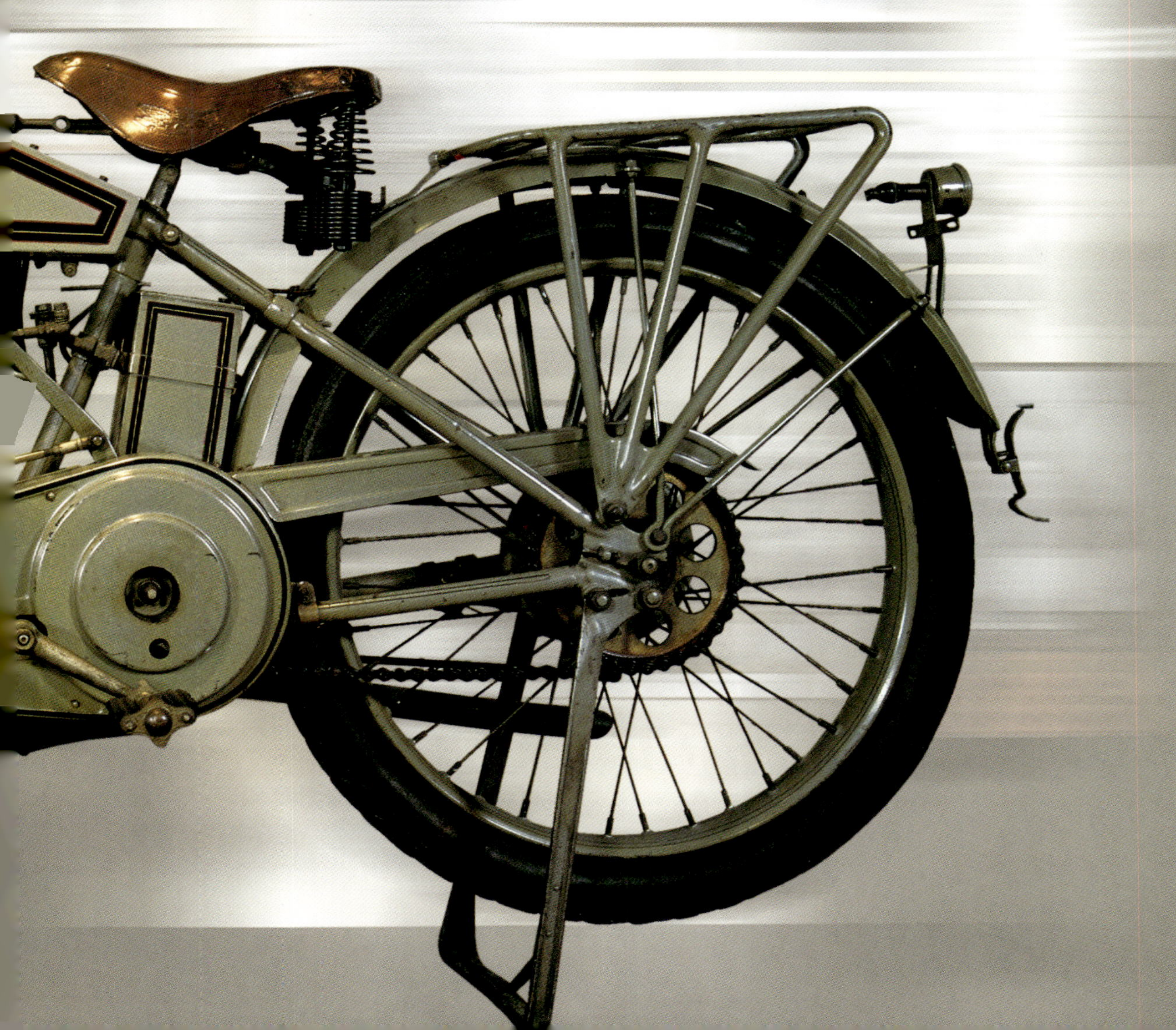

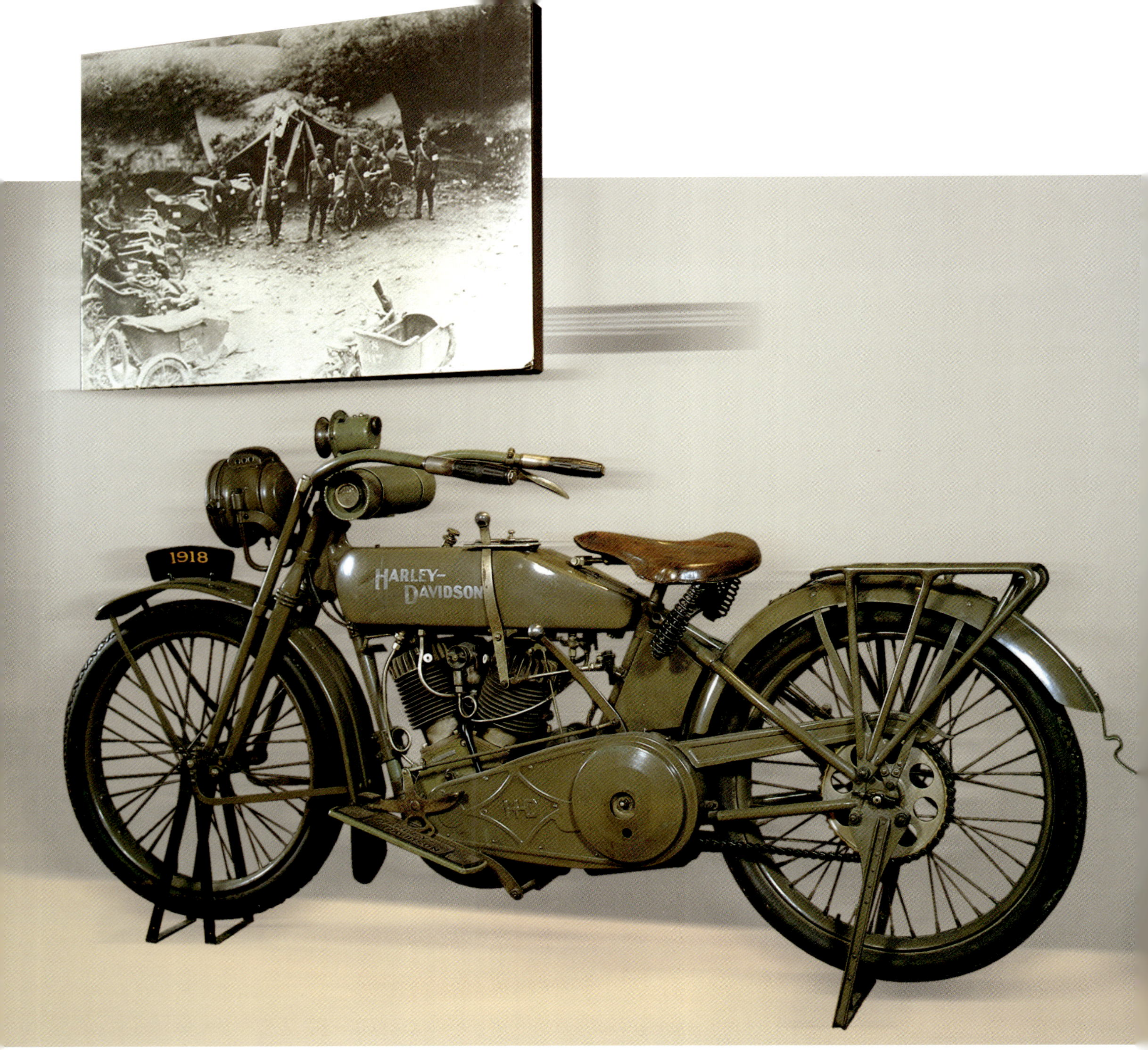

142 The 1918 two-cylinder J model was the world's first motorcycle to be used in military conflict. The gas bottle, located between the two handlebar grips, powered the headlight and horn. Equipped with a sidecar, this type of motorcycle was driven to Germany by Corporal Roy Holtz on November 12, 1918. It was propelled by a 7-horsepower 988.83 cc displacement V_Twin engine, with a three-speed gearbox and simple train gearing. Changing gears meant disengaging the clutch. The transmission was based on a roller chain. Primary and secondary drive chains, covered by sealed drawn steel guards, could be adjusted separately. With its automatic carburetor and throttle, the V-Twin started easily. Three carburetor fittings ensured a better mixture at every speed. Drum brakes worked like internally expanding brake shoes, worked by a pedal on the right footrest. The brake drum had an 18.85 cm diameter and was 2.3 cm wide. The fuel tank held over two gallons; the auxiliary tank held three-quarters of a gallon and the oil tank held four quarts. A curved steel frame, especially designed and reinforced to allow the attachment of a sidecar, supported the whole.

143 The Sport Twin model was a real novelty for the Harley-Davidson Company. The 35.6 cubic-inch model was introduced in mid-1919, and was provided with a 6-horsepower motor that could achieve 45 to 50 miles per hour. Over the first few years of production, it sold for $335, but four years later the price went down to $275. Thanks to a crankshaft counterweight, this two-horizontal-cylinder engine could operate at 4500 rpm. It was fitted with ball bearings for the main engine bearings and the connecting rod small end bearings. It had a magneto flywheel housed in a sealed case. The oil level in the gearbox was kept constant, thanks to an automatic lubrication system. The transmission design, which included lubrication by oil steam coming from the crankcase, was supposed to ensure longer chain life with less maintenance.
A high-voltage magneto mounted on the top of the crankcase powered the ignition and was easily accessible. This model was only manufactured from 1919 to 1922, because Harley-Davidson found that it didn't win enough customers away from the Indian V-Twin Scout.

144-145 This machine, introduced in 1922 under the model names 22 F and 22 J, lasted for six years of production. The electrical systems of the two models differed: 22 F was fitted with a magneto and 22 J had dynamo. Available with engines of 61 and 74 cubic inches, the motorcycle sold for as much as $390. This machine was powerful enough to be combined with a sidecar. Its gas consumption rate was reasonable, as a gallon of gasoline would run sixty miles. This low gas consumption ensured it a good fuel distance thanks to its 3-1/2-gallon gas tank. It had a four-quart oil tank and good front and rear lighting systems, with a six-volt battery positioned towards the rear. An aluminum gearbox and pistons made for better heat dissipation and greater power. In 1928, a front brake introduced in addition to the existing rear brake, was located at the right foot. The decorations on this model echoed the graphics of the first Harley-Davidsons. During the 1920s, design improvements reduced engine vibrations by half, thus improving the V-Twin's duration. In 1924, aluminum alloy pistons were introduced, developed after lengthy testing, which meant even further improvements in performance.

These lighter pistons, four rings, and drilled connecting rods lightened the moving parts and made the engine more powerful. Since these engine parts were easier to cool, too, high speeds could be maintained over longer distances without danger of overheating. Cooling did not demand too much oil, and the newly designed rings, pistons, and axles resisted wear longer than the old cast-iron parts, thus increasing reliability.

In 1924, both machines – those fitted with old pistons and those with new – were offered in the Harley-Davidson catalog. Models with aluminum pistons could achieve higher speeds than those with cast-iron pistons: ten miles per hour faster solo and twenty with a sidecar. As for lubrication, an Alemite system efficiently allowed the hand pump to send the lubricant at a 500-pound pressure to the desired point. In 1924, the exhaust doubled its volume and the kickstart system was improved. Motorcycles with electric systems had a 6-volt generator, an accumulator, a headlight, a taillight, an electric horn, and a hand switch with an automatic alarm signal; fuses, housed in a box, protected these new systems. These motorcycles were painted olive-green and decorated with broad maroon stripes and gold pinstripes. An ammeter, speedometer, auxiliary brake, and luggage rack were available as optional parts. The handlebar was manufactured from a single piece of tube with a diameter of an inch, ending in rubber grips. The grips housed the controls, operated by double steel wires sheathed and hidden in the handlebar.

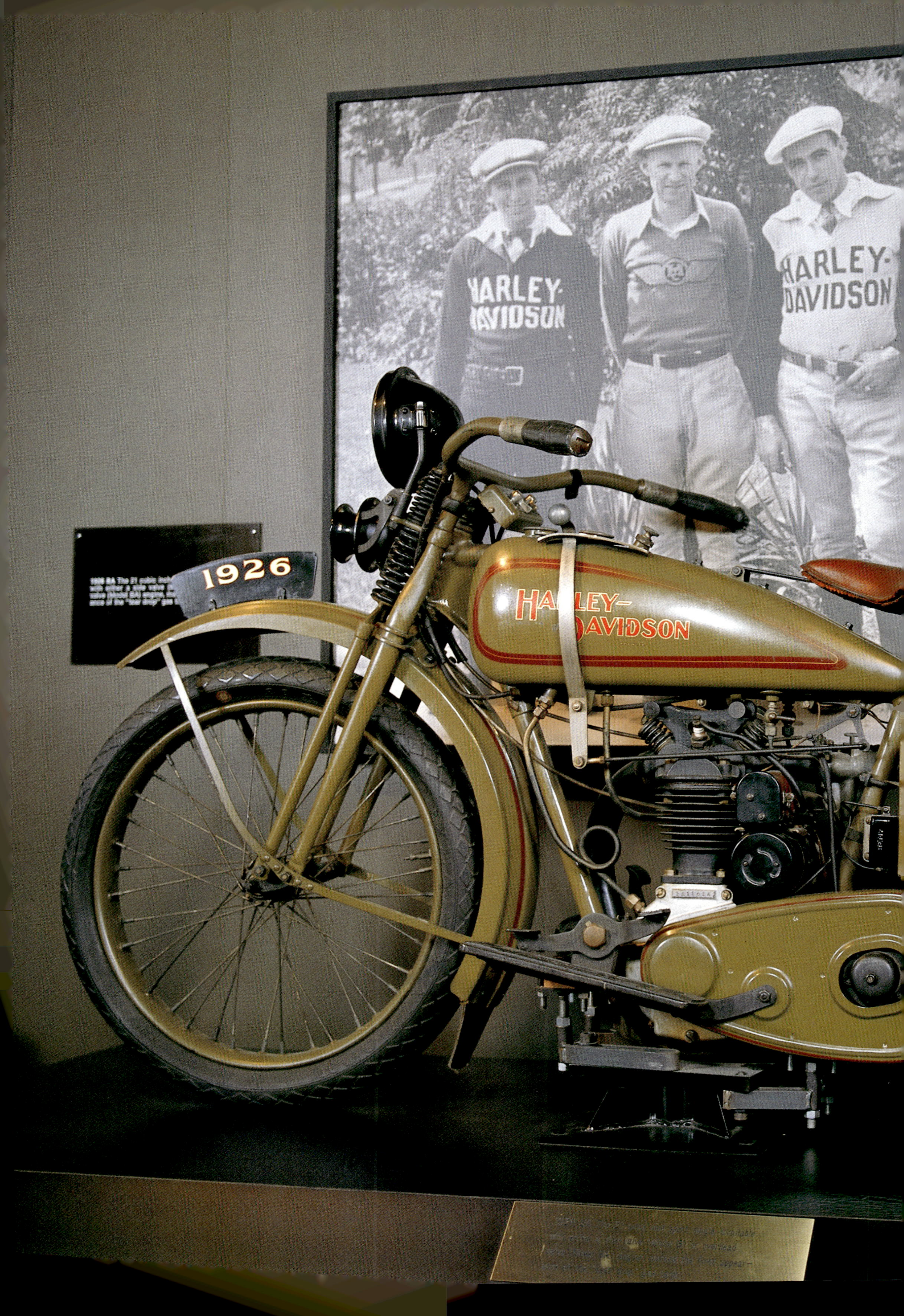
1926
HARLEY-
DAVIDSON

146-147 A new motorcycle came out in 1929, 23 years after the birth of the Silent Gray Fellow. Nicknamed the "Peashooter", it was lightweight and had a 21 cubic-inch (350 cc) displacement and a Schebler carburetor. The engine was lubricated by a mechanical oil pump. A rich mixture control, located on the right side, made starting easier. The wheels were 55 cm. This motorcycle was equipped with a tool kit, including tools for repairing tires.

It was manufactured for ten years with two types of engines available: one with side valves and a capacity of eight horsepower, and the other with twelve-horsepower head valves, priced between $210 and $275. It could be fitted with either an electrical system or simply with a magneto. Like the 1919 Sport Twin Model, it became fairly successful in Europe, New Zealand, and Australia.

In 1935 Joe Petrali (who, a few years later, was to establish a world record on a Harley-Davidson powered by a V-Twin) won several dirt-track races with this little machine. His thirteen victories prompted the American Motorcycling Association to create a new motorcycle competition category.

During the last years of production, this motorcycle's power was increased, so that ultimately it could reach over 60 mph.

148 The 45 cubic-inch, three-speed Model D, produced from 1929 to 1931, was the first 750 cc displacement Flathead. This machine, fitted with a medium-displacement two-cylinder engine mounted on a frame like those of one-cylinder engines, was conceived to compete with Indian middle-of-the-range motorcycles. Its appearance was characterized by two headlights on the 1929 and 1930 models. Some motorcycles had a metal cylinder for a toolbox fixed under the lens. The 750 cc Flathead engine with side valves had cooling fins on each cylinder head cover. The W series faced a slight problem with the three-speed gearbox, which shifted into neutral between second and third gear. Despite this problem, the 45 cubic inch proved to be an excellent engine.
In 1931, starting with this machine, Harley-Davidson built a prototype called "Bullet" that could reach a speed of 85 mph. The strength of the low-compression 45 cubic-inch Flathead made it possible to run on both regular and super gasoline. Over 22 years this engine proved to be successful on all counts – from city roads, used by police departments to rough terrain, used by the Army during World War II. It was a solid, reliable engine that required little maintenance.

***148-149 and 149 bottom** The 1932 V and VL models, produced for five years, had a 74 cubic-inch displacement (1200 cc) and were integrated with the 80 cubic-inch (1340 cc) model in 1935. Despite the introduction of the Knucklehead, these side-valve machines continued to be popular, renamed UL or ULH, in 74 and 80 cubic inches respectively. These were Harley-Davidson's most popular engines, closely related to the 45 cubic-inch models with which they shared some engine parts. Their production ended with the introduction of the Panhead in 1948. The tires on models UL and ULH were identical. They were designed to allow rapid disassembling. Once the Knucklehead came on the market, these machines adopted similar bodywork and logos.*

150-151 The 1930 V and VL models which became models U and UL in 1937, still used with 74 cubic-inch (1200 cc) Flathead engines. There was also an 80 cubic-inch UHS model, provided with a special trunk for sidecars. (This model had existed previously as a low-compression version, under the reference VSD). All these Harley-Davidson Big Twins fitted with Flathead engines look the same as those fitted with Knuckleheads. Only the engines help differentiate the motorcycles.

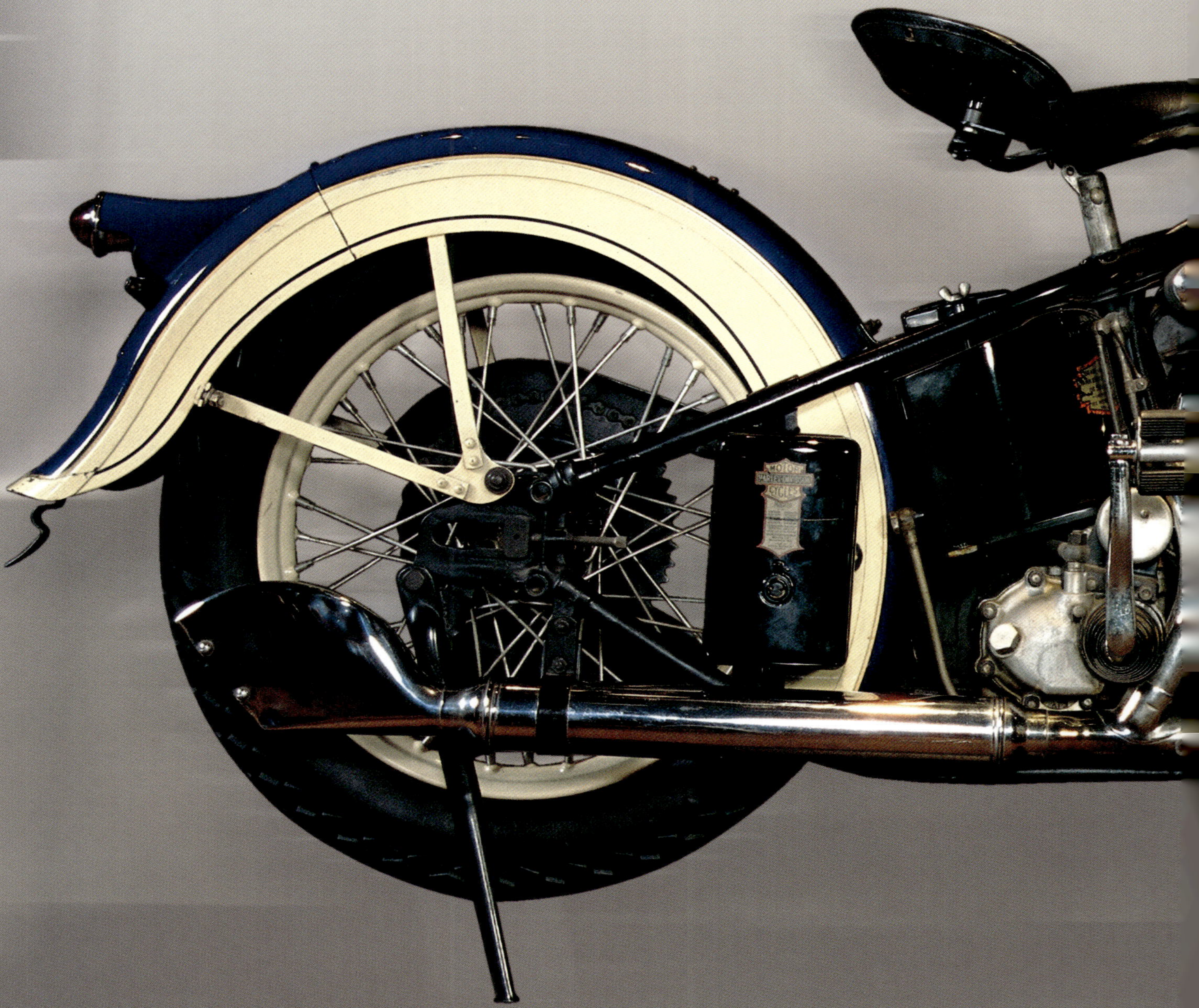

1936

1939
HARLEY-DAVIDSON

152-153 The Knucklehead, the first overhead valve V-Twin model, was called Model E when it first appeared in 1936. It had a 61 cubic-inch displacement (1000 cc), and the special shape of its cylinder head covers made it easy to recognize. Early on in production, it faced some lubrication problems. The oil didn't flow in the correct direction, flooding certain parts of the engine and leaving other gears without lubrication.
Harley-Davidson corrected this flaw on the 1937-38 models. Three steel friction clutch discs were adapted on all models, followed by a larger air filter; a reinforced rear braking system (which increased braking power by 44 percent); better contact with the brake lever, no longer made of aluminum; and, especially, a new speedometer support inserted in the middle of the tank, which is still present in today's Harley-Davidsons. The gear-change lever and the clutch pedal were better positioned.
The early lubrication problems kept the Knucklehead from selling at first. Sales dropped during the war, then afterwards started soaring.

154-155 The engine fitted on the WLA model, manufactured from 1940 to 1945, actually started as a V-Twin Flathead 45 cubic-inch side engine in 1929. During World War II, this model was called "Liberator," and it certainly appeared as such to the French. After the war, the V-Twin lasted for a long time, used for civilian purposes until 1951, when the K series replaced it.
This engine proved to be comfortable, solid, reliable, never causing maintenance problems and always easy to adjust. This Flathead was available with 61 cubic-inch (1000 cc), 74 cubic-inch (1200 cc), or 80 cubic-inch (1340 cc) displacements.
In 1929, the first Harley-Davidson equipped with a 45 cubic-inch Flathead engine appeared: the three-speed Model D. In 1931, this engine was fitted on the Model D, and on the DL and DLD (both sport); it was called R from 1934 to 1936 and then W through 1951. Its toughest competitor was the 61 cubic-inch (1000 cc) Knucklehead, launched by Harley-Davidson in 1936 which sold with great success after the war, thanks to its overhead valves with rocker arms.
When Harley-Davidson finally decided to stop manufacturing the 45 cubic-inch side engine, war had been declared. Americans involved this model in the conflict, where it served particularly well. By the end of 1943, Harley-Davidson had produced some 40,000

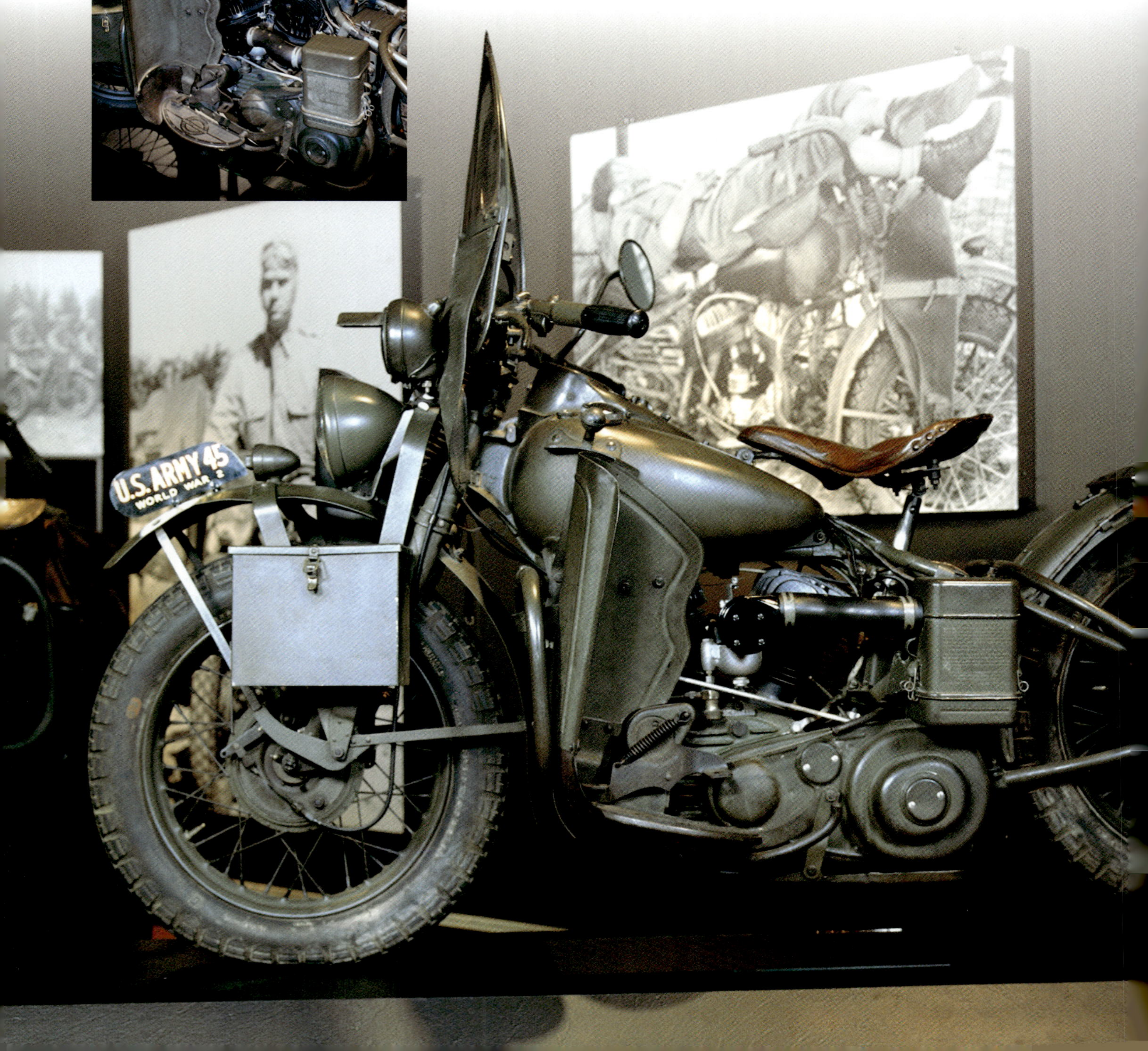

WLAs for the U.S. Army. Originally, this motorcycle was fitted with an ammunition box and a submachine gun holster on each side of the front fork.
The WLA cylinders were made of cast iron, with valves placed at the side, to the right of each cylinder. Valves were operated by four cams, one for each. A pump fixed at the end of the crankshaft ensured dry-sump lubrication; the exhaust system was a two-in-one. The fuel tank consisted of two sections, as do most Harley-Davidsons today. But at that time, if the left side contained fuel, the right contained oil. The grip located on the right end of the large handlebar housed the timing advance control. (On the Harley-Davidson WLCs for the Canadian Army, the system was placed on the left.) The clutch was no longer controlled by a lever on the left of the handlebar; it was located next to the left footrest instead. The driver changed gears in the classic way, with a control to the left of the tank. A fairly simple dashboard included two odometers (one of them a trip odometer), an oil pressure warning indicator, a fuel level indicator, and on-off indicators for ignition and lighting. An engine guard protected the crankcase.

156 The 750 cc XA was produced for only one year (in 1942) at the request of the U.S. Army, which had been impressed by the performances of the German Army's Zündapps and BMWs. To manufacture this motorcycle, the Harley-Davidson company just copied a German machine with a flat two-cylinder engine, converting dimensions to inches. In fulfillment of its Army contract, Harley-Davidson manufactured one thousand machines, intended primarily for North Africa, where the shaft transmission, unlike the WLA chain transmission, could withstand the sand.

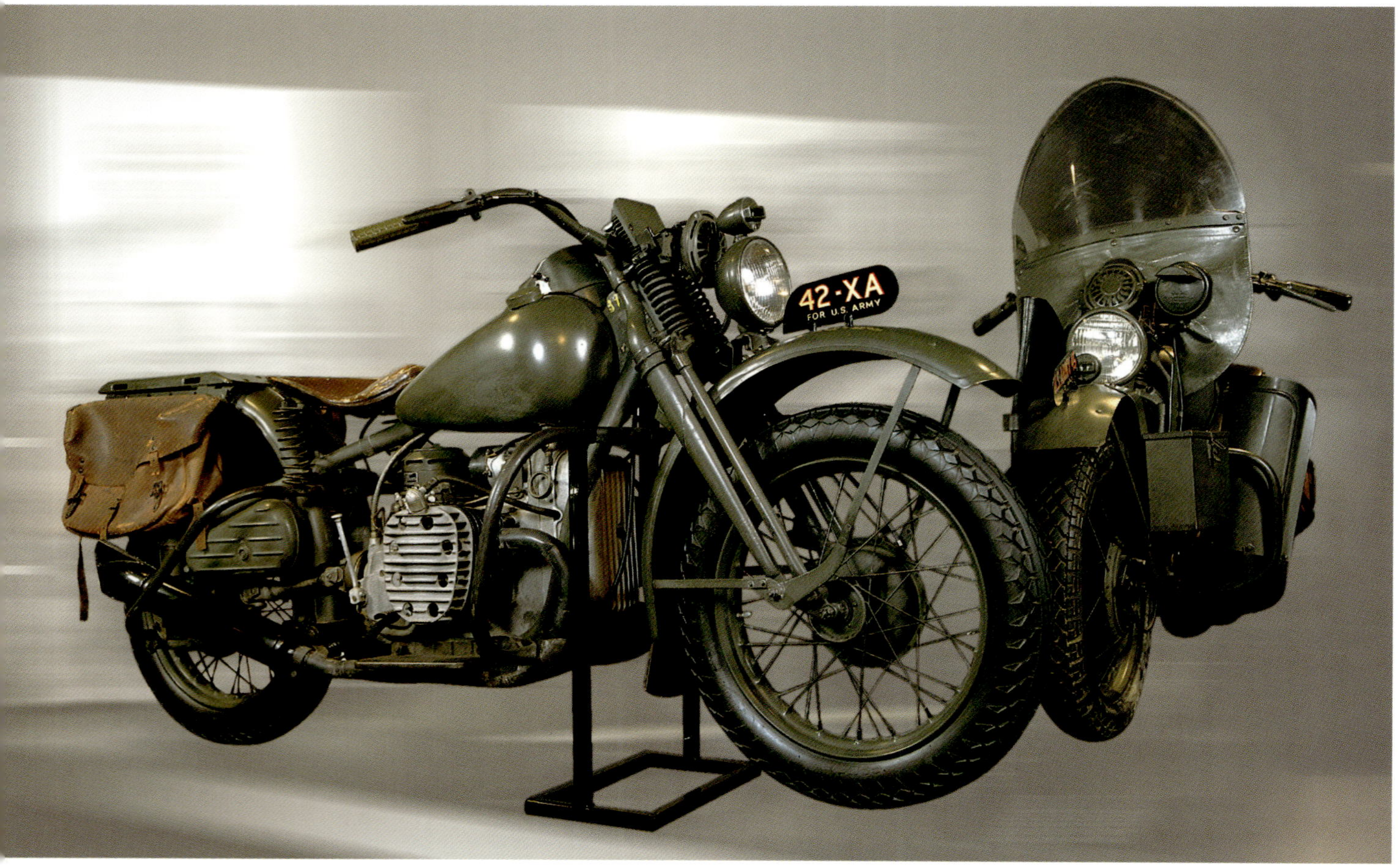

157 Just before the war, which slowed down production and sales of the Knucklehead, its displacement was increased up to 74 cubic inches (1200 cc). The bigger engine was called FL in 1941. Meanwhile, the EL (with 1000 cc) was thriving. In 1940, a combined total of 4,069 Knucklehead EL and ES models were sold; in 1941, 5,149 Knucklehead EL, ES, FL, and FS models were sold. In 1942, sales fell to 1,743, and to 200 in 1943. In 1944, the sales curves started up again, with the sale of 535 Knuckleheads, but truly positive figures (6,746 sales) only began in 1946. After that, ten thousand Knuckleheads were sold every year until 1948, when this model was replaced by the Panhead.

158-159 In 1948, the Panhead, a new generation of engines, replaced the Knucklehead, establishing itself firmly in one arm of the Harley-Davidson line with displacements of 61 cubic inches (1000 cc) and 74 cubic inches (1200 cc). From 1949, the Panhead equipped the new E Series called Hydra-Glide. This engine still had valves in the cylinder head, but they were made of aluminum, to ensure better cooling. In 1954, the 61 cubic-inch version disappeared. The Panhead was manufactured for 17 years as part of the Duo-Glide and then the Electra Glide lines. 1965 (the only year in which an Electra Glide was manufactured with a Panhead engine), was the last year of production for this engine. The Hydra-Glide, Duo-Glide, and Electra Glide were equipped with hydraulic telescopic forks. These models are still remembered for their harmoniously gentle and rounded features and their timeless look. Today, the Heritage Softail Classic and, even more so, the Springer Heritage are developments of the Hydra-Glide and the Panhead, which may account for their commercial success.
Besides aluminum rocker arm covers, the Panhead was fitted with hydraulic tappets for the rocker arm rods. Thanks to these

aluminum parts, the cylinder heads dissipated heat more rapidly, reducing noise. A new camshaft was designed and a new oil pump was installed for better lubrication. In 1948, a few technical improvements were brought to this model, notably a corrosion-proofing painting, a more comfortable seat, and a pale blue color, unusual for a motorcycle. When a telescopic fork was added to the Hydra-Glide, the front drum-brake power increased by 34 percent. In 1952, the hand-operated gear change became a left-foot-operated gear change, with clutch disengagement on the handlebar. In 1953, the figures of the odometer, housed in the tank, appeared in miles, with the numerals 1 to 12 representing miles per hour from 10 to 120. It was an excellent year for this motorcycle and its V-Twin, highlighted by a golden V on the front mudguard, turned silvery the following year. In 1954, the Hydra-Glide was fitted with a chrome-plated chain guard, a horizontal taillight and, as an option, a couple of rigid bags on the back of the motorcycle.

160-161 After 1947, Harley-Davidson faced fearsome competition from English brands such as Triumph and Norton. The English motorcycles appealed to Americans, thanks to their sporty look and technical innovations. In 1952, Harley-Davidson adapted the English-style foot gear-changing system, incorporating it into the Model K. This bike, with its new 45 cubic-inch engine replacing the WL Flathead engine in 1952, was supposed to give Harley-Davidson an edge over the British competition. At first its structure required some improvement to become really reliable and efficient. Its WL heritage (side valves, identical bore, and stroke), its development over the years (55 cubic inches in 1954), and its variety of sport models allowed Harley-Davidson to accumulate victories for several years. On the front, it had a hydraulic telescopic fork at the level of the shock absorbers, and, with a weight of 450 pounds and a five-gallon gas tank its 30-horsepower engine provided good flexibility at low speed.
A slightly improved racing model had a 40-horsepower engine. In 1954, the KH was presented with an 883 cc (55 cubic-inch) engine, but this model was not very successful and disappeared after three years. Harley-Davidson built various specialty racing versions of the Model K, including: the 1953 KK, a specialty Model; the KHRM, which first appeared in 1954; the KRTT,

another racer model with powerful brakes and reinforced shock absorbers; and the KR, a racing motorcycle manufactured for 17 years (1952-1969), which took part in C-class races. This model, in all its versions, represents an ancestor of the Sportster, first manufactured in 1957.

1957
SPORTSTER
HARLEY-DAVIDSON
SPORTSTER

GEN

162-163 In 1957, American youths, put off by the problems of the 1952 Model K, were drawn by the superior performance of British motorcycles. So Harley-Davidson presented the XL, which kept the frame and low engine of the K models. The valves were no longer at the side but the engine was still cast iron. The exact displacement was 883 cc (54 cubic-inches), but was called 900 cc. Several options were offered the first year, including a double seat, a windshield, and saddlebags. During the first year, 1,983 Sportsters were manufactured, and by 1959, Sportster production jumped up to 2,053 machines, spurred by the success of Models XLH and XLCH, which were introduced in 1958. Some internal modifications were introduced, such as the adaptation of polished valves, but magneto lighting was maintained, resulting in rather poor reliability. Because of its small tank size, this model was called the "Peanut."

At that time, California retailers urged factory managers to create a sport model based on the Sportster; as a result, the XLCH was launched in 1958 and became street-legal the following year, when Harley-Davidson manufactured 1,059 XLCHs. Sportster production nearly doubled by 1961. During these years, new shock absorbers were added along with a new range of colors and a new streamlined look, called "jet-stream styling." Some parts were manufactured in aluminum. In 1964, the braking system was improved and a new 12-volt electrical system was developed. In 1965, the Harley-Davidson Motor Company signed its last contract with the U.S. Army, supplying 1000 XLAs (i.e. Sportsters) equipped to meet military needs. In 1967, an electric starter was fitted on the XLH, and a kick was all that was needed to start the XLCH.

***164** In 1958, the Hydra-Glide was replaced by the Duo-Glide, the motorbike of the 1960s. It was fitted with a rear hydraulic brake and a neutral point indicator light in the counter unit. In 1963, the front brake drums were made of aluminum. Little by little, the Duo-Glide was improved. It became a superb motorcycle, with features still found in some of today's models. Harley-Davidson first produced bikes with the Panhead in 1948 (the E Series). The last Harley-Davidson bike with a Panhead engine was the 1965 Electra Glide, in its first year of production. For the rest of its career, the Electra Glide had a Shovelhead engine. The Electra Glide, equipped with a push-button electric starter on the handlebar and a five-gallon gas tank, was exactly the type of motorcycle needed for highway driving, and it met consumer demand perfectly. Comfortable, reliable, and tough, it became the benchmark for those who liked to cover long distances on a prestigious motorcycle.*

164-165 The Sprint Model C was a 250 cc machine developed in cooperation with Aermacchi. Produced from 1961 to 1974 with 250 cc, then from 1969 on with 350 cc, it was a Harley-Davidson by name only. In spite of being manufactured by Aermacchi with quite a dubious look for a Harley-Davidson, it became fairly successful within its market niche during the first ten years it was sold in the United States. Then its future became more uncertain, as Japanese manufacturing techniques invaded the market.

Beginning in 1960, for five years, a scooter called the Topper was also produced, referred to as A and AU. In all, 5,246 such motorcycles came off production lines – not a large volume. A horizontal cylinder, a two-stroke engine, and a 1.7 gallon tank were its main features.

The Topper competed with the Italian Vespa, which, in the 1960s, was one of the most popular means of transport with certain drivers, such as students and artists. A small trunk under the seat of this small machine held books or other objects. In 1942, this model, with five to nine horsepower, sold for $445.

Aermacchi and Harley-Davidson produced a 250 cc four-stroke Trial model, the Sprint H, in 1962, to allow race drivers to stand out in competition. Subsequently, Harley-Davidson provided it with a lighting system so that it could be driven in town.

The company continued developing the 250 cc Sprint C over the years. In 1964, it was fitted with an Italian carburetor, made by Dell'Orto. In 1967, its name changed to 250 Sprint SS, and, in 1969, it became the 350 cc Sprint SS. In 1972, it was finally called SX 350, manufactured until 1974.

166 Within the old Harley-Davidson range, there was a one-cylinder moped called the M 50, manufactured in cooperation with Aermacchi between 1965 and 1968. It was a carefully conceived and finely decorated two-stroke model, targeted at customers who were not particularly interested in the performance of the big Harley-Davidson models. One of the two models launched on the market had a character and look that recalled the excitement of competitions. The M 50 cost $250; the sport version, $275. To compete against companies supplying more modern models, in 1970 Harley-Davidson turned these models into the M 65 and the M 65 Sport, sold at more affordable prices. Between 1965 and 1972, AMF/Harley-Davidson produced 44,955 M-style mopeds, including the M 50, the M 50 S, the M 65, and the M 65 S. The last year of their production under these names was 1972; after that, the moped was renamed 65 cc Shortster and, from 1973 until the end of 1974, when production stopped, it was referred to as Z90 and X90.

167 In 1969, Harley-Davidson merged with AMF, an association that lasted for more than a decade. The Sportster benefited from the merger: production grew to 7,800 items in 1969, and to 10,775 in 1970. In 1970 and 1971, the Sportster came with an optional "boat-tail" back. This machine and the Super Glide, equipped with a similar back, were the first two customized motorcycles created by Harley-Davidson. Moderately successful in their own time, they are in great demand today. At that time, customers could also opt for a front mudguard made of fiberglass rather than sheet metal or for a small "Peanut" tank, rather than a large one. Fiberglass boat tails were produced in Tomahawk, based on a design by Willie G. Davidson. But the line was not as successful as expected and fewer than 3,000 XLHs were manufactured, as compared with the 5,500 XLCHs produced that year. The majority of these motorcycles were assembled with conventional mudguards, unlike the first Super Glide which, in 1971, was supplied exclusively with the boat tail. The main mechanical changes on the 1971 Sportsters are easy to sum up. The clutch system was replaced by an oil-bath clutch and, since stronger pressure on the disks was then needed, the springs were adjusted more rigidly, making the hand control less flexible. Platinum-plated screws and a condenser came in the timing case, to make the ignition system more damp-proof. This machine was rare not only due to its boat tail, but also because it was the last cast-iron engine supplied with 900 cc. In 1972, all Sportsters were supplied with 61 cubic-inch (1000 cc) engines in an attempt to become more like their Japanese and English competitors.

168-169 and 170-171 The Shovelhead engine appeared in 1966, replacing the Panhead on the Electra Glide. This engine provided about 15 percent more power, and it was hardly improved beyond that over the 18 years of its production. It became synonymous with Harley-Davidson's lack of brand reliability during the 1970s, in the period when AMF increased production and eroded product quality, which caused considerable economic difficulties for the company.
The Shovelhead was developed due to the fight for survival against Japanese competition. A well-assembled Shovelhead, with quality components, was more reliable than the previous Harley-Davidson Big Twin models (the Knucklehead and the Panhead), even though it can't compare to today's Evolution models. The last Shovelheads were manufactured in 1983 and sold in 1984, when the Blockhead (Evolution) engine came onto the market.
Besides the Electra Glide, the Shovelhead appeared on other renowned motorcycles, shaping styles that still appear today.
The Wide Glides, the Low Riders, and the Super Glides – all were produced with this engine – which still has an important place in the spirit and history of Harley-Davidson.

1200

POLICE

Harley-Davidson
XR-1000

HARLEY-
DAVIDSON

172-173, 174-175 and 176-177

As a result of the merger between AMF and Harley-Davidson, Sportster production continued to grow, going from 8,560 motorcycles in 1970 to 10,325 in 1971 and 23,830 in 1974. Sportsters accounted for more than two-thirds of the overall company production, which included seven different models. Technical improvements on the Sportster or XLH came steadily. In 1970, Harley-Davidson replaced the electric kick-start with a battery-coil ignition. Purchasers could choose between a small and a large gas tank. In 1971, the oil bath clutch was replaced by a dry clutch, with the resulting greater plate pressure meaning stiffer springs and hand controls. Japanese competitors became all the more a threat, appealing to American customers in price, models, and technology. Sportster displacement went up to 1000 cc in 1972. In the same year, two other models in that line, the XLH and the XLCH, were equipped with a long, uncomfortable seat and a small tank. Factories produced 7,500 XLHs and 10,650 XLCHs. Before 1981, apart from rare exceptions like the Café Racer, Sportsters had only changed aesthetically. In that year, Harley-Davidson produced 41,606 motorcycles, 10,102 of which were Sportsters. In 1982, Harley-Davidson released a limited series of the Sportster to celebrate its 25th anniversary; in 1987, it did the same for its 30th. In 1983, two special models, equipped with an oval air filter, came out: the XR 1000, as a limited series, and the XLX. The XLX, a stripped-down, less expensive version of the Sporster, was their top-seller that year, with 4,892 sold. The XR 1000 was not as successful, selling a total of 1,018. Nevertheless, the following year, the company changed its design and reissued the same model, fitted with a lighter and more rigid frame, a four-gallon tank, a halogen light, and a new battery.

This extraordinary motorcycle kept evolving. In 1984, the front wheel was designed as a single disk of larger diameter, the clutch became more flexible on certain models, and an alternator replaced the generator. The Sportster preserves this look today.

HARL
DAVIDSON
1000
GOODYEA
EAGLE A/T

GOODYEAR
EAGLE A/T

Harley-Davidson
1000

178-179 Among the various models associated over 36 years with the important Sportster line, the XLCR Café Racer is a landmark in the history of the Harley-Davidson Motor Company. Released in 1977, a total of 3,124 of these bikes were manufactured, selling for $3595. Even though this model enjoyed modest commercial success, in aesthetic terms it was too advanced for its time. Now, collectors are constantly hunting for it, and its popularity has pushed its value much higher. Further from ordinary motorcycles than most of the Harleys, this model, the product of Willie G. Davidson's creative imagination, was probably drawn from the famous English Café Racers of the 1960s, yet it is no mere imitation.
The engine is that of an XLCH, slightly improved and mounted on a duplex cradle frame, which was derived from the famous racing model, XR 750. The same frame served other Sportster models from 1979 on. Its caster angle was 29.35°, and rear suspension was secured by three adjustable dampers. A classic telescopic fork supported two ten-inch disks, braked by new hydraulic calipers with simple pistons. The back drum, which still equipped the Sportster XLH and XLCH 1000 in 1977, was replaced with the same type of disk as in the front. Seven-spoke wheels made of a light alloy were meant to increase the Café Racer's sporty look. The position of the driver, determined by the small, flat handlebar; the controls, positioned further back; and the black elements in the chassis all highlighted this motorcycle's aggressive look. A small fork-head fairing, vaguely reminiscent of the Japanese style, ensured the rider's comfort. A tank solo-seat unit, with a four-gallon capacity, was added to the extension.
All mechanical parts were black, including the twin exhaust pipe with mufflers on each side. A Keihin throttle carburetor with a return pump ensured the fuel supply of this 500-pound machine. With the high 9-to-1 compression rate on the XL and XLCH 1000 engines, the bore and stroke remained unchanged and the rebore dimensions produced the exact displacement of 61 cubic-inches (997.5 cc). In 1977, this Café Racer, running 61 horsepower at 6200 revolutions per minute and reaching a manufacturer's top speed of 120 miles per hour, was fast enough.
But the Café Racer was not a commercial success. In 1977, 1,923 Café Racers came out of the Milwaukee factory. In the following year, production dropped to 1,201. In 1979, only nine Café Racers were manufactured.

1000
GOODYEAR
EAGLE A/T

180-181 The year 1986 marked the debut of the Evolution engine in the XLH Sportster series, taking the place of the OHV V-Twin, which had powered the series since 1957. In 1988, the 1100 was replaced by a 74 cubic-inch (1200 cc) version. The 1986 Sportsters appeared in either a 883 or 1100 cc version. From then on, the Sportster Evolution has enjoyed a brilliant career. It was available in

two versions (883 and 1200 cc). Since 1986, it has undergone some interesting technical improvements, such as the addition of a fifth gear in 1991 and a standard toothed belt to replace the transmission chain, which had already been abandoned on the high-displacement motorcycles (1340 cc/80 ci).
In the middle of the 1990s, the Sportster family was aimed at two different customer niches: sportspeople and traditionalists.
The Sportster XL 1200 had a new three-gallon tank, found also on the XL 883 Standard and XL 883 Hugger. The Hugger remained the same for several years, despite modifications carried out on all the other Sportster models.
The XLH 1200 S had features that made it even more sporty. The front and rear suspensions were entirely adjustable. Even the spring pre-load and the hydraulics could be adjusted. There was a cartridge front fork and the rear gas dampers had an attached tank. Floating double disks ensured front braking, and thirteen-spoke wheels, made of aluminum, were fitted with soft rubber tires. A flat track handlebar combined with a more sporty seat.
The XL 1200 C stood out with lots of chrome, notably a chrome-plated headlight without a cover. The handlebar fasteners, the risers, and other parts were also chrome-plated. The engine block was black and chrome, compared with the polished aluminum plating in other Sportster engines. The front wheel was a 21-inch model, as on the 1340 models with a chopper look; the rear wheel had a full rim, similar to that of the Bad Boy. The speedometer was electronic, the tank logo was made of metal, and the seat was about two inches lower than on the old Sportster 1200 cc. On this model, front braking was limited to a simple disk, floating back and forth, like that fitted on the Big Twin customs.

Harley-Davidson
883

182-183 What do people expect from Harley-Davidsons?
That they remain faithful to their heritage, yet improve over the years. This means that the motorcycles should preserve their classic design and character, that they should develop without changing radically, that they should be gradually perfected for quality and comfort.
Judging by sales, the most popular machines are the Softail models. These now offer electronic speedometers, a one-piece console, and a new ignition switch with an anti-theft key. The 2012 Softail line includes five models: Softail Blackline, Fat Boy Lo, Fat Boy, Heritage Classic, and Softail Deluxe.
The 2012 Dyna line includes the Dyna Super Glide Custom, Dyna Street Bob, Dyna Switchback, Dyna Fat Bob and Dyna Wide Glide. All have benefited from a frame modification since 2006, lowering the seat about an inch. A new battery support and electric housing make these motorcycles easier to maintain.
The 2012 touring range, grouped together with the term FLT, includes the Road King, the Road King Classic, the Electra Glide Classic, the Electra Glide Ultra Classic, the Electra Ultra Limited, the Street Glide, the Road Glide Ultra, and the Road Glide Custom. If the driver could choose between a carburetor or an electronic injection for several models until 2006, since 2007 all the Harley-Davidson are with injection. The electronic injection systems are particularly reliable. Furthermore, they permit easy cold starts, regular idling, and less pollution. The 1997 machines fitted with injection have a troubleshooting system on the dashboard to make maintenance easier. The fairing has also been changed to make motorcycle maintenance easier, reducing the number of assembly parts from 42 to 14; moreover, internal fastening screws for the fairing have been eliminated, so that it is no longer necessary to dismantle the headlight to remove the external fairing. Similarly, windshield fasteners have been reduced from seven to three and are no longer visible from the driver's seat. A new, easy-to-remove cover, located over the main switch block and protecting the cruise control, the loudspeaker, and the accessory circuit switches, makes it easier to adjust the main switch block and the handlebar.

EERING

184-185 The Evolution Softail marked the revival of the Harley-Davidson brand from late 1983 on. It was fitted with a modernized version of the traditional V-Twin engine, made of aluminum and mounted on a sprung frame that evoked the old rigid frame – just what tens of thousands of Harleyists deeply desired. In 1986, Harley-Davidson took an active role in the celebration of the 100th anniversary of the Statue of Liberty by organizing a marathon. The race started from numerous cities and ended at the foot of the statue, with profits dedicated to the statue restoration fund. The company also sold limited-series Liberty edition motorcycles, a total of 1,714 units in 1986, and for each motorcycle sold, $100 was allocated to the restoration fund.

Harley-Davidson's emphasis
nological improvements is
d by an emphasis on the rich
of the past. America's only
g motorcycle manufacturer
ost sight of what it has taken
this far and
ing into the
1986 FLHTC ELECTRA GLIDE
LIBERTY EDITION
HARLEY-DAVIDSON
Classic

***186-187** When, in 1984, the Evolution engine, or Blockhead, appeared, they were fitted on all the FXRs, on two of the three FLs, and on the 1985 FXST. Only one Electra Glide and the Super Glide were still equipped with the Shovelhead engine, and that just for one more year. The FXR line, positioned as the most modern of motorcycles and the most futuristic made by Harley-Davidson, grew to include a light tourism version, the Sport Glide FXRT, in 1985. This machine was equipped with a profiled fork head, saddlebags, and even a small trunk on some models. It aimed at competing with the Gold Wing motorcycles, but it only managed to appeal to public servants, especially to the police. Harley-Davidson withdrew it from the catalog in 1993, after the public had neglected it for years. The FXR line, less popular with customers than the Softail, was also gradually abandoned during the first half of the 1990s, replaced by the more modern Dyna range, drawn directly from the 1991 Sturgis FXDB Dyna and commemorating the 50th anniversary of the show in 1990.*

Harley Davidson

From the Chopper to the Harley-Davidson Customization

Birth of the Chopper Concept

Initially, creating a chopper meant highlighting the engine and its performance, mainly by working on the finish, the paints, and then increasingly the frame. Over the years, the finishing work – accessory research and paint frenzies – gained importance until today they are an end in themselves.

But today, as a sort of return to the past, after an era of choppers based on all brands of motorcycle engines, Harley-Davidsons are back in the lead. At present, with the new cruisers fitted with V-Twin engines and impressive new Royal Star or FC6 machines, choppers can mean customization based on motorcycles of any brand.

A high-performance chopper

The name "chopper" originally described a motorcycle cut into pieces like a chunk of meat with a thick, sharp blade. At the end of the 1940s and the beginning of the 1950s, there was no efficient way to supercharge a road motorcycle. The only efficient parts were destined for the racetrack, in racing Harley-Davidsons and other motorcycles built specifically for racing; but they were too unreliable for the road. For someone who wanted racetrack performance on the road, the best solution was simply to adjust the weight. Since American motorcycles of that time had been adapted for touring over long distances and were equipped for comfort, they carried lots of extras, such as saddlebags, a windshield, huge exhaust systems, large wraparound mudguards, and huge seats, for example. One gained twenty or thirty pounds by eliminating all these accessories. A smaller seat was easy to find. You could remove the huge exhausts and replace them with parts developed for the competition sector or, even better, manufacture your own. You could take away the huge original tires, recovering the hubs or not, and fit smaller wheels in their place, both on the back and the front (especially on the front). At the same time, the too-large GT handlebar could be replaced, as could the huge tank. The name that emerged to describe the resulting bizarre motorcycles was "chopper."

By the 1960s this had been refined to an art. Here's how one contemporary article put it: "Apart from rare exceptions, the wild biker drives a Harley 74 which originally weighs 800 pounds and which can be brought down to 550, once it has been stripped, transformed, and made lighter. The chopper then consists of a massive frame, a tiny seat and a huge 1200 cc engine, twice the size of the English machines such as the Triumph Bonneville, the BSA Lightning Rocket, motorcycles whose speed approaches 120 mph. The original Harley-Davidsons were not up to such performances and needed to be transformed to become competitive against their rivals, driven by standard riders, and against the souped-up cars and motorcycles used by policemen. In order to make a Harley 74 grant the same performance, the first priority is to change the power-to-weight ratio, which explains why the wild bikers strip their motorcycles as much as possible and only keep the essential parts, sometimes removing even the front brake. Once the motorcycle has been stripped it can already ensure completely different performance. But, since often this still isn't enough to leave the other motorcycles and the cops behind, the engines are also reworked, with performance camshafts, modified valves and a higher stroke-to-bore ratio. Only the compulsory accessories remain, such as the headlight, the taillight and the rear mirror, sometimes reduced to a simple dentist mirror in compliance with the law. The weight advantage can still be increased by halving the tank, removing the front mudguard, truncating the rear mudguard at the top of the wheel, adapting a large-horn handlebar, reducing the seat, lengthening the forks, adding a clutch pedal called a 'suicide pedal,' adapting megaphone-like mufflers, a tiny cyclo-like front wheel, a front headlight reduced to the minimum, chrome plating and painting, etc.

"A chopper is often a veritable work of art with an impressive resale price, not taking the hours of work into account. Lighter, highly chrome-plated, and covered with complicated paintwork consisting of many overlapping

coats, the chopper is a superb and distinguished machine, so mechanically perfect that it is difficult to imagine it on the highway, running at top speed, driven by a brute in a dubious state just before crashing against a tree or a car . . ."

188-189 The chopper, idealized through its role in the film* Easy Rider*, was built from a Panhead engine and a rigid frame. Its look influenced generations to come.

189 At first, the traditional chopper was designed with a minimum number of elements, a big handlebar, a lengthened fork, and no front mudguard.

Chopper transformations

Nothing remains unchanged, especially in a field where creation prevails. Over the years, the chopper developed quickly, integrating many new ideas. The people riding and transforming their motorcycles were individuals with differing tastes and aspirations. Everybody started personalizing and modifying his machine in his own way, with each resulting motorcycle ending up far from the original.

It is worth remembering that only American machines developed in this way, as the Japanese were not yet present on the American market and the English machines, already simple and bare, did not yield to such radical transformations. It was a logical development, stemming from the need to make the large American motorcycles as efficient as the English ones. At that time, English motorcycles were agile, light, and powerful, due to their low weight. Their mechanical conception was also more modern than the big American machines. So, to counter the English and satisfy the chopper fans, Harley-Davidson put a more modern version of the side engine 750 on the market; it had a gearbox integrated in the engine sump, a double cradle frame, and a telescopic fork. But this engine was painfully lacking in power, compared with either English machines or the stripped-down big Harleys. This motorcycle was replaced by the KH 900 in 1954, an 883 cc engine with 38 horsepower, 5000 rpm, and a manufacturer's top speed of 100 mph. Those who liked performance and stripped-down machines finally had a motorcycle that lived up to their dreams.

Over the years, however, the big Harley-Davidson 74 machines were chopperized more often than other models. The Panhead lives on, now more than ever, synonymous with the chopper in bikers' minds. It was recognized as the best motorcycle of its time, not only by independent bikers but also by the Harleyists who belonged to official clubs and wanted a reliable, faithful motorcycle to cover long distances.

This phase of motorcycle evolution owed a lot to the media, too. The film *The Wild One,* starring Marlon Brando and Lee Marvin, brought American biker lore to Europe. This film retold, in romanticized Hollywood style, the story of Hollister, a little town in California (located more than fifty miles south of San Francisco, near Gilroy), where a biker revolt had taken place in 1947. During an Independence Day celebration, which included a motorcycle race, about four thousand motorcyclists, spectators and racers descended on Hollister. Some racers came with official teams but some were independents, competing even though they were not members of the American Motorcycle Association. The A.M.A., organizing the race, wanted to exclude the nonmembers. Between the numbers of bikers who had been let down and the quantity of alcohol consumed with meals (which actually dried up all the bars), the party deteriorated. The press, with nothing else to report at the time, jumped on the situation, quickly turning the motorcycle phenomenon into bands of terrible bikers roaring around on motorcycles and plundering the country. A few more meaningless episodes brought fresh interest, and the press embellished the stories, keeping up pressure in the biker-hooligan vein. The A.M.A. suddenly pronounced itself in favor of kind bikers – those who only used their motorcycle for sport or for the weekend family drive, whether associated with the A.M.A. or not. On the other hand, the association blamed wild bikers for all the scandal, identifying them as members of renegade clubs, not associated with the

A.M.A., accounting for only one percent of all motorcyclists . . . to which the birth of the phrase "one percent" is related.

The Wild One retold the Hollister story. After the release of the film in 1954, a number of wild biker clubs, issuing their own sets of rules, sprung up all over America. Rejected by conformist society as well as by other bikers, they decided to live their life according to their own aspirations, astride their choppers and stripped-down Harleys. The performance they got out of transformed motorcycles proved to be vital, but the aesthetic side of the chopper also started to take shape. With the passing years, the rush for performance gradually lost ground in favor of the quest for fashion. The chopper was finally tamed, and even turned into a real industry by certain dealers.

Polite Chopper

Once choppers had their own specialized shops and custom motorcycle builders, the search for a different look began. This occurred thanks to the gradual association between motorcycles and West Coast hippies. Performance – the original reason for creating these machines – was gradually forgotten and the look became vital. 1968 proved to be a particularly eventful year in the motorcycle world.

That year, Honda presented its four-cylinder CB 750 in the U.S. A little earlier in the same year, Triumph put the Trident on the market. And, also in 1968, *Easy Rider* came to the screen, starring Peter Fonda and Dennis Hopper in their search for freedom on two magnificent choppers. *Easy Rider* inspired the motorcycle paintwork of a whole generation, and those styles still inspire certain customized paintwork today.

The Harley-Davidson motorcycle benefited most from this media attention because this motorcycle, always considered a comfortable machine, familiar by its association with policemen, now took on a new look. The Harley-Davidson police version did not disappear, but everybody became aware that it was possible to drive a Harley in a different way. The brand was reborn, through stripped-down cycles and original choppers – machines now synonymous with freedom and rebellion. It was a fantastic feeling . . . even though much of the image was untrue. But that image took hold, and it is still alive today.

***190** Today, the Evolution engine is most widely used for choppers. Harley-Davidson has been selling machines fitted with this engine since 1984. But the recipe for building a traditional chopper is still the same: rigid frame, long fork, big handlebar, and the minimum possible equipment.*

***190-191** Fairly new choppers, based on old Harley-Davidson Shovelhead engines, can even be found in Europe. They are particularly popular among purist enthusiasts.*

The Dream and the Media

192 top left and bottom The cult film **Easy Rider** *made stars Peter Fonda and Dennis Hopper the idols of several generations of Harleyists.*

192 top right ZZ Top took part in the ninetieth anniversary celebrations, leading the parade with their Harley-Davidson Hogzilla, their Cadillac Cadzilla, and a chopperized Harley-Davidson still under construction.

192-193 top **The Wild One** *was the first film to put motorcycles center-stage. The motorcycles in the film were not all Harley-Davidsons, though, as shown in this photo of Marlon Brando on a Triumph Speed Twin.*

192-193 bottom There are several exact replicas of the motorcycle driven by Peter Fonda in **Easy Rider**. *At big meetings, one such replica escorts the truck from "Easyriders," an American retail shop.*

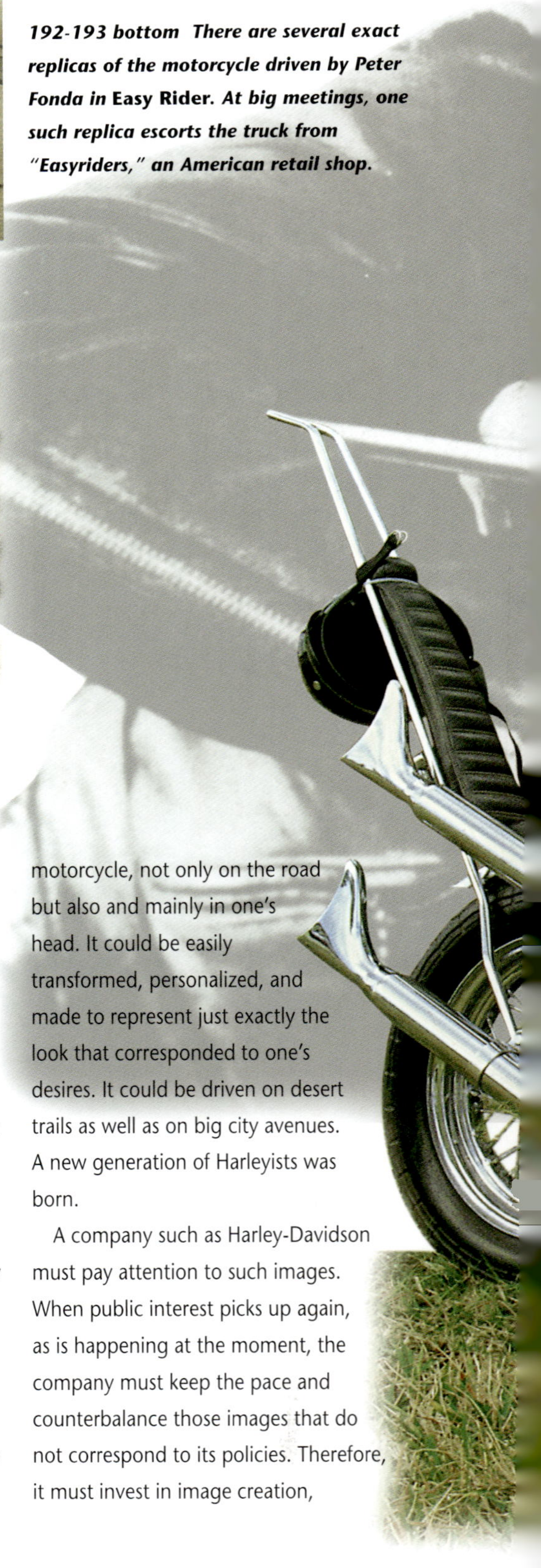

It was not easy, though, for Harley-Davidson to counter the image held over from the 1947 events in Hollister, California. When the gathering erupted into a gang scene, the town became terrified and the situation which could have been solved calmly, escalated. Policemen from throughout the region were called in to restore order. *Life* magazine reported the event with a picture taken by Barney Petersen, portraying a drunk biker on a Harley-Davidson, towering over a stack of beer bottles, a bottle in each hand. The situation fanned the flames of polemic, and wild bikers and one percent clubs reacted, calling themselves the "Booze Fighters," "Satan's Slaves," "Satan's Sinners," "Satan's Daughters," and the "Winos." In July 1972, *Life* published another article, but by this time, many other magazines had run stories about the "wild bikers."

As a result of all this media attention, the Harley-Davidson became a symbol of virility. Those who drove Harleys were terrifying tough guys with searing eyes. The films of the '60s portrayed this rather idiotic stereotype. Luckily, in 1969, *Easy Rider* revealed a different side of the Harley-Davidson. In it, the two lonely heros, travelling with an alcoholic lawyer (Jack Nicholson), discover the wide open spaces of America, as well as the hostility of some of its rural citizens. Their violent end evokes sympathy, not scorn.

The film combined hippies, drugs, and the chopper, opening new horizons for bikers and especially for future Harleyists of the time. The Harley-Davidson became an escapist motorcycle, not only on the road but also and mainly in one's head. It could be easily transformed, personalized, and made to represent just exactly the look that corresponded to one's desires. It could be driven on desert trails as well as on big city avenues. A new generation of Harleyists was born.

A company such as Harley-Davidson must pay attention to such images. When public interest picks up again, as is happening at the moment, the company must keep the pace and counterbalance those images that do not correspond to its policies. Therefore, it must invest in image creation,

arranging media events to convey a positive look. In Harley-Davidson's case, these efforts were accomplished through the company's association with the musical group ZZ Top.

In 1991, ZZ Top took part in the Bike Week at Daytona Beach, celebrating the 50th anniversary of the event. Special Harleys were of course on hand: two customized Fat Boys and the Cadzilla, a 1948 Cadillac with an entirely new body, designed to match the Harley-Davidson look. Peter Chapouris transformed the two motorcycles, taking three months to cover them completely in aluminum. He redesigned the whole look, fitting the headlight with a fork head and transforming the back of the two bikes to resemble a Cadillac. Billy Gibbons, guitar player and singer of ZZ Top, spoke of how their passions were motorcycles, cars, and good-looking girls. It all worked so well that afterwards, ZZ Top signed a contract with Harley-Davidson.

When the company celebrated its 90th anniversary, ZZ Top led the parade. Tens of thousands of Harleyists were there for the great event, all with their dream vehicles. The Hogzilla motorcycle, the Hogzilla car, and the Eliminator car all were there, plus new motorcycles, never before seen. Media efforts like these made people forget the films of the '60s . . . except for *Easy Rider,* which continues to thrive as a cult film, and is now a touchstone for several generations of Harleyists.

A Way of Life

An excerpt from *Male,* a paper from the 1960s that explored the spread of the chopper phenomenon throughout the United States, offers an idea of the image of bikers at that time, which irresistibly evokes the one with which we are familiar from films: "A very hot day in 1954, a tall, bearded, and tanned guy stops his Harley with a controlled sideslip in front of the bar favored by the bikers of an American city. His faded denim shirt, with sleeves cut in one whole piece, shows the colors of a club. Suddenly, with a turn of the grip, he tears the air of the road, on a quiet Sunday afternoon, steering his 5-feet high handlebar, disclosing his soaking sweaty armpits, before shutting off the gas.

"He leans his bike on its stand, polishes the glittering chrome of its forks, which are much longer than the standard ones, with a torn handkerchief. Then he gives a look around and nonchalantly dries his greasy hands on his jeans, encrusted with oil. This is the real thing." According to various sources, the choppers

appeared most often in California after World War II. No other state combined so many favorable factors for the development of such an original vehicle – the permanent sun shining on varied scenery, from the mountains to the desert, the large cultivated fields (instead of huge cities), the more than 600 miles of coast, the easy money, the challenge of heavy fines given for speeding by muscular cops riding huge Harley-Davidsons, and especially the constantly growing population, due not only to births but also and mainly to immigration from other states (and from abroad).

In the United States, as elsewhere in the world, young people began to distrust adults, or worse yet, despise them as being powerless. Many teenagers felt that they were being forced to conform to a narrow-minded society with puritanical customs, and bred to conform politically. The protests and the reactions grew as the generation gap widened. Young people let off steam with cars, sometimes taking to violent enter-tainment. These games, such as being the last driver to jump out of a car running at full speed toward the edge of a cliff, could have fatal consequences. The concept of a gang increasingly came into play with these games. Not only is union strength, but also unusual and therefore disquieting actions – for example, wearing sunglasses and leather jackets with gang colors, and driving big noisy motorcycles – inspired fear in outsiders. These gangs gravitated naturally to the unique chopper look and the dissatisfaction with the status quo which it represented.

Today, the Harley-Davidson is still a way of life for an increasing number of people, but these ways of life have diversified. The number of people actually living as the original "wild bikers" is small in comparison with the number of Harley enthusiasts simply living with their passion for the wonderful mechanics of the Harley-Davidson. The latter gather together with their friends or within Harley-Davidson clubs, H.O.G. or not. The Harley-Davidson clothes and accessories allow anybody to wear the colors of the company, whether or not they own a motorcycle, because fashion has become part of the Harley-Davidson phenomenon. As a result, many fans who do not yet own a Harley-Davidson still follow the rhythm of the "belle" of Milwaukee, waiting for the happy day when they will finally own the magic V-Twin; while others still simply wear all the clothes and accessories as a fashion statement. Unfortunately, the latter group will disappear as soon as the phenomenon fades . . . unless they become seduced by the machines themselves in the meantime.

***194 top** Whether based on a Panhead or on a Shovelhead engine, whether their frames are rigid or springy, whether they were simply derived from Softails with modified frames and a lengthened fork, choppers still symbolize the same spirit of freedom.*

***194 center and bottom** New custom styles and new choppers have continued to appear over time. Some of them evoke the old machines. Others have been simply stripped to enhance their performances. Always, care for detail has been the priority.*

***194-195 and 195 top** Beginning as rough, stripped, and lightened machines, designed simply to increase performance, choppers have gradually developed a look which is as sophisticated as ever.*

Chopper Evolutions

As the evolution of the chopper progressed, specialists started to establish thriving shops in the United States. The quest for a unique look began to prevail over the practical aspects of the motorcycle. The vital importance of the engine was gradually forgotten, and chopper design went crazy. Whether it was propelled by a V-Twin Harley-Davidson, a Japanese 4-cylinder on line, or an English Twin became of little importance to the user. The chrome

plating, the excessive sissy bars, the endless forks, the exhausts outrageously raised or with twisted pipes, the scrap-iron flower and arabesque patterns, sometimes added later, made everybody forget the origin of the chopper – which was connected with the quest for performance.

In the United States, mail order of chopper parts grew, allowing everybody to build his own personalized motorcycle at home. This industry, which began in the mid-'60s, has thrived over the years. But as the bikers' aesthetic quest became increasingly specific, mail order alone no longer sufficed. A real industry with stores and workshops has developed for manufacturing and marketing parts, not only for motorcycle transformations but, gradually, for their entire manufacture as well.

The chopper underwent a spectacular development but preserved an image connected with its troubled past: newspaper articles continued to promptly report any accident involving one. Subsequently, certain states passed laws designed to limit chopper eccentricity. The best example is the laws enacted in Florida, which regulated the maximum distance between the handlebar grips and the gas tank, and called for exhaust systems of a certain shape. These laws were passed in the '70s, when the competition and the search for originality made manufacturers build increasingly souped-up machines, which sometimes became dangerous to drive. The Show-Bikes – motorcycles just for exhibition – made their first appearance at the same time. Only the engine worked, so that the machine could cover the few yards needed to be classified as a motorcycle. The Show-Bike never ventured onto the road, not only because of complete law infringement, but mainly because the machine simply could not be driven. As a result of these laws and the settlement of dealers, individual D.I.Y., which often proved dangerous due to lack of knowledge disappeared in favor of safer and more reliable work. (It must be admitted that accidents arising from unskilled, amateur work were the impetus which led to the enacting of the most restrictive laws.) Of course, not every individual initiative disappeared, but those who did not feel able to perform adjustments to their frame or fork appealed to dealers or to friends who had the necessary knowledge,

196-197 ***The Harley-Davidsons of the Softail type are an excellent example of chopper evolution, although this model only appeared in late 1983. Once sporting a rigid frame look, this machine was first fitted out in the spirit of traditional choppers, with a long fork and a big handlebar. Later, as its look moved closer to the Low Rider, it became more efficient.***

instead of risking their lives on machines that might thereby endanger the chopper phenomenon. In the '70s, the Harley-Davidson was gradually surpassed by the Japanese machines, not only on the road but also in the chopper field. Japanese four-cylinder engines increasingly replaced the big V-Twin in frames specifically designed to house them, or in original frames, such as those with rigid backs, which had been suitably modified. The multitude of available possibilities made it necessary for the choppers to be classified into categories, but these were not fixed and overlapped according to the modifications implemented on each model. In general, the Street-Racers can be defined as stripped, slender, thin choppers offering good performance and aimed mostly at attaining this goal. It is from this type of chopper that the Café Racer, a name which was even adopted by standard motorcycle manufacturers, later differentiated itself. At present, the traditional chopper can be defined as a Chop fitted with a mainly rigid, long and slender frame, but with a big enough caster angle to house a fork which is longer than the original, a small tank, a flat and small or backward and long or raised handlebar, a seat which suits the spirit of the machine, and a sissy bar. There is great freedom concerning the Chop, as the very principle at the basis of this machine implies. Then there is the Low Rider, a long, low machine, with or without an inclined steering stem and with an original or shortened fork. The Low Rider is the result of extensive work on details and on engine performance, and it is more compact. The Street-Drag, on the other hand, appeared with the shape of a Street-Racer towards the end of the '70s and the beginning of the '80s. It is a machine designed for high performance, with a modified engine, low and compact lines, and sometimes the addition of N.O.S. in order to improve its starting efficiency after stopping at traffic lights. The Show-Bike is a motorcycle built for exhibition, not for driving.

The Oldies are recent machines that try to resemble the old models. The chopper has a bright future, as it is supported by the custom field, where crazy transformations are more and more applied to original Harley-Davidsons. This original approach looks to the chopper.

The Old Dream Models

The ancestors are pampered, restored, and often customized according to the rules of the time. These ancient motorcycles, which are either the products of American brands which have disappeared or of Harley-Davidson, make up part of America's cultural heritage.

The Harley-Davidsons are those which most often seduce the collectors, but it is also true that they are easier to restore than the majority of the other, vanished brands. It is possible to find parts for old Harley-Davidsons in the catalogs of some big companies that specialize in supplying parts which can be adapted to modern bikes. This is not the case for other brands.

198 Although a few Flatheads still exist together with the few remaining Knuckleheads, discovering an XA of this type is exceptional. Not only are the military versions of this machine very rare, since only a thousand units were built, but it is practically impossible to find a machine of this type that was transformed for civil uses.

199 The Sportster can be adapted for customization in a variety of ways. Whether it becomes a rigid-frame chopper or a performance model, it is equally popular as both the old models and the recent Evolution versions.

Sportsters

The work of customizing and transforming Sportsters grows, both in the United States and in Europe. The choice of the Big Twins as a base for transformation prevails, but the Sportster, which sometimes becomes the second motorcycle for Big Twin owners, benefits from the advantage of time. In other words, not only are there an increasing number of elements which can be used for modifying the Sportster, but because many Sportster owners also own another bike, they don't hesitate to put it out of operation for a while, during which they can create something remarkable.

The Sportster's classic lines are often changed radically, and it is possible to come across Custom Sportsters with a Big Twin line, or even a Low Rider or a Softail one, with a Fat Boy tank, or suitable mudguards – not to mention the growing use of Softail frames manufactured by specialized factories. However, motorcycles with the look of old machines, of choppers, and of touring models, which still show a quest for performance, are also very common.

Sportster and Performance

For Americans, Sportsters are the ideal base for building very sporty motorcycles. The Buells are the best example of this idea, but they are not the only ones to occupy this performance niche. There are both mechanical and aesthetic preparation kits. They can generally be found at shows and, on rare occasions, on the road. According to the owners, this engine is full of potential and unexploited possibilities and requires a minimum amount of work to grant the better performances than the Harley-Davidson Big Twins.

200 These machines, based on modified Sportsters, have a sleek look even though they are intended mainly for shows. In the dragster performance field, old engines are also very popular.

***201** From an aesthetic point of view, choppers based on Knucklehead engines provide the best design, but they are difficult to find. The Panhead, also quite elegant, is more common and has been customized as well.*

Show Panheads and Knuckleheads

The beautiful choppers of the past, whether they come from radically transformed machines or from machines that have simply been stripped and fitted with personalized parts, always elicit a nostalgic sigh. These machines, unlike others from the same era which have simply been restored, have a special soul. They are the owners' unique creations and, even more, they represent the first steps towards the chopper and towards customization. It was these bikes which were first modified by Harley enthusiasts who desired more efficient and aesthetically unique machines.

The more restrained chopper option uses a Knucklehead or Panhead engine on a rigid frame, with a long fork of the "oldstyle." Panheads are also good for mounting in the Low Rider style, to increase performance. Beyond their technological suitability, both engines are also suited aesthetically to this type of transformation. Certain people have even tried to fit out Evolution engines to make them look like Panhead or Knucklehead engines, by using cylinder head covers and an X-Zotic crankshaft cover.

Shovelheads

The Shovelhead engine has a past which is rich in history, though this is not always to its advantage. It is possible today to drive a Harley-Davidson equipped with a perfectly restored Shovelhead engine. Of course, this engine will never be as reliable as an Evolution model or reach similar levels of performance, but there is still pleasure to be taken from this fantastically souped-up, die-cast engine.

The Shovelhead is always present at the shows that take place during the great meetings. The many types of transformations on display are the result of research aimed either at improving performance or finding the Harley-Davidson look of the future.

Luckily, the traditional chopper category with a rigid frame has a very strong presence as well, but it is no longer the only category in which motorcycles equipped with Shovelhead engines take prizes.

***202 top and bottom** The Shovelhead engine appeals to many Harleyists as a motorcycle that reconciles reliability, performance, and a look which is both different from the Blockhead engine and return to the always fashionable retro motorcycle look.*

***202 center** Some Shovelheads are turned into high-performance models by being fitted with compressors and internal parts taken directly from dragster racing bikes. Several bikes with engines of this type still take part in races against modern aluminum engines.*

Harley-Davidson rats

Shows mainly include Harley-Davidsons that have been customized or chopperized, but other types of vehicles are also welcome. They may be Harley-Davidsons or a different brand, and may be equipped with two, three, or occasionally even more wheels. Among the many categories, the most peculiar may be the one including motorcycles referred to as "Rats." A "Rat" is a machine that looks like a motorcycle which has not been maintained: it is dirty and covered in all sorts of accessories, some of which are only for decoration. Although some "Rats" are well enough maintained to be able to cover a fair number of miles without hitches, despite the numerous oil leaks which are intentionally left unrepaired, some are real rolling wrecks.

"Rats" like those of Big Daddy Rat have a strong presence at the shows they attend, where they generally appear laden with accessories, gadgets and, obviously, the oil and dirt which they have accumulated over the years, honing their look little by little.

203 The most well-known owner of a Rat motorcycle fitted with a Harley-Davidson engine is Smithy. He travels and regularly wins races on his Rat Knucklehead.

Low Riders and Dyna Glides

Low Riders and Dyna Glides are very often used as bases for customization, as they were the first machines to benefit from the initial transformations carried out by dealers.

Low Rider engines were isolated from the frames and consequently spread fewer destructive vibrations than the Softails. Even so, the original look of the Low Rider did not appeal to the public, and so it was less appreciated than the Softtail. This explains why bargain Low Riders are much cheaper to buy than the Softails. Because the Low Rider Evolution frame was deemed aesthetically unappealing, Arlen Ness decided to create a more refined one, returning slightly to the design of Low Rider frames which were equipped with old Shovelheads. In the end, Ness's line prevailed.

Later, other types of transformations appeared which led to today's mature version: performance is still always a priority, as the Low Rider frame can house any improvement designed to produce more efficient machines which also hold the road well. But the quest for modern version of the chopper is also a recurring feature. Thus, there is a growing trend towards unique, sophisticated models, where the Low Rider frame is hard to see.

After the new Dyna Glide range replaced this type of frame, the approach remained the same, and the work carried out has had fairly similar aesthetic results, apart from an oil tank integrated around the gearbox. This development should permit much more adventurous research.

204 Low Riders were still largely present at shows, even after the arrival of the Dyna Glide. Dynas rapidly replaced them because of the greater possibilities for customization that they offered.

Softails

The Softail frame provides the widest range of possible transformations. Originality and aesthetic research prevail, but technical detail is never left aside. There are many kinds of V-Twin engine blocks mounted on Softail frames, whether they are Harley-Davidson originals or not. Similarly, there are many Softail frame manufactures for Big Twin Evolution engines, in which it is possible to adapt non-Evolution engines, such as the Shovelheads, the Panheads and the Sportsters engines, and everything may either be fitted as rigid on the frame or on Silentblocs.

As for the style, many different looks are popular. There is the nostalgic look of the old Duo Glide, which is very popular because it is easy to obtain from Softail Heritages; there is the radical chopper look obtained through a modified and lengthened frame and an outsize fork; there are the customized machines which are obtained through chrome plating and Hi-Tech parts, sometimes flabby, or fitted out; and finally, there are motorcycles which have a look aimed at reaching high performances. In short, the Softail is suitable for any kind of customization and chopperization.

***205** Softails were customized in the old chopper style, then transformed into efficient, aggressive, and tightly packed models. Later, their lines were lengthened for a more modern look.*

***206-207** The frames of the Softtail motorcycles have been modified by lengthening and refining the frame, or even almost completely replacing it with parts available on the market; the end result bears little resemblance to the original.*

206 bottom *This Canadian Softail is decorated with paintwork over tinted chrome: a new technique which allows paint to adhere to the chrome and offers new possibilities to the fans of customization.*

207 bottom *This Softail represents a streamlined modification. The tank is longer and sleeker, and the body, including the articulated front of the oscillating arm, is thinner, for an overall high-performance look.*

Road Kings

The Road King is a touring machine which has only been available in the United States since the end of 1994; nevertheless, it is often modified and benefits from rather daring customizing. A craze for this motorcycle has been developing amongst Harley-Davidson enthusiasts, and the trend towards customization is reinforcing the phenomenon. Although a certain amount of personalizing is obtained through paintwork, this is not always the case. Accessories are also used and radical transformations are done on frames, tanks, and even saddlebags. Once the saddlebags have been modified, plates are sometimes integrated, and then the motorcycle is lowered. The result is a touring motorcycle, which has been personalized, but retains the advantages linked to its original purpose: comfort over long distances. Owners refine the line of the motorcycle, but not at the expense of losing its practical attributes; in other words, it keeps certain pieces of equipment which, in another time, would have been removed to achieve radical transformation.

***208 top** Despite modern aluminum parts, this classic Road King customization has a nostalgic character, thanks especially to the fringes.*

***208 center and bottom** The Coca-Cola Road King represents the quintessential customized look. Every bit of bodywork, down to the tiniest detail, has been modified to evoke the look of the Coca-Cola logo. This motorcycle called "Secret Formula," represents the harmonious efforts of Skip Hoagland's team.*

Electra Glides

Although the Electra Glides are beautiful machines for travelling and have always attended all the Harley-Davidson meetings in large numbers, they do not have a comparatively strong presence at custom shows. This is due to the fact that, as this motorcycle is mainly intended for tourism, it is customized less frequently.

Customization is often confined to specialized paintwork, chrome plating, and accessories. It is very difficult to implement more radical transformations on a motorcycle of this type and preserve its functional character at the same time. Luckily, this fact does not stop many very beautiful versions from being produced and it has the advantage of making them stand out among a crowd of other Electra Glides. The most current transformations are the lowering of the motorcycle, the integration of the saddlebag into the overall look of the machine, and the reduction of the fairing; the picture is completed by the addition of details softening the general line of the vehicle. The result speaks for itself, and Electra Glides of this type are the stuff of dreams, even if they do function less well.

209 Since the Electra Glide is a popular touring machine for families, its customization often involves just adding accessories. Some people push further, though, entirely disassembling the machine to add a personal touch.

A Unique Motorcycle for Everybody

Some Specialized Companies which Moved the Custom Bikes from Handicraft to Industry

The companies specializing in supplying parts and accessories for the Harley-Davidson motorcycles have multiplied, reaching amazing international sizes with yearly turnovers of several million dollars. This success has been achieved by devoting themselves to selling parts for a single motorcycle brand: Harley-Davidson. After a modest start for most of them during the '70s and the opening of a few others during the '80s and the '90s, they began building a good reputation amongst specialist customizers, as well as Harley-Davidson dealers. Custom Chrome Incorporation, Drag Specialties, Chrome Specialties, Nempco, Mid U.S.A., and dozens of others distribute parts for these legendary motorcycles. These companies are so successful that many of them now manufacture their own parts for Harley-Davidsons as well, and promote them exclusively in their catalogs. Since the middle of the '90s, the Harley-Davidson company has reacted strongly, hoping to gradually regain their share of this lucrative market.

Custom Chrome

The biggest company among those specializing in aftermarket parts is called Custom Chrome and has a financial statement showing an increase of 16.2% over the first 4 months in 1995. The Custom Chrome company was founded in 1970 and it has become the most important distributor of parts for Harley-Davidson motorcycles in the world, with a general headquarters in Morgan Hill, California and distribution offices in Visalia, California, in Louisville, Kentucky and in Harrisburg, Pennsylvania. Custom Chrome Incorporation perfectly illustrates the above-mentioned phenomenon concerning the distributor who starts manufacturing his own parts, which he sells in addition to all the official products and parts for Harley-Davidsons listed in most of its catalogs. These parts are manufactured by companies owned, in this case, by C.C.I., such as RevTech, Premium, Dyno Power, and C.C. Rider. Every year, Custom Chrome presents new motorcycles customized with its unique parts and proposes styles which are sophisticated enough to make Harley owners feel like transforming their motorcycles.

This policy has spread among big distributors, but the people who increasingly work this way – that is to say, who present the largest number of motorcycles with radically different styles – are called Drag Specialties.

***210 top** This special commemorative machine was built by John Reed, who managed the Custom Chrome research and development department. This Softail, a lottery prize at Sturgis' fiftieth anniversary party in 1990, was later bought back by C.C.I. from the happy winner.*

211 top This Custom Chrome Softail's tank is decorated with paintwork portraying the company truck used during the mid-1990s for special events and showroom appearances.

211 center The considerable number of elements available means Evolution, Shovelhead, or even Panhead engines may be used.

211 bottom The old-style chopper, based on Softail models, is still in demand among Harley-Davidson aficionadoes. Because of its popularity, many parts are available for those who want to modify the cycles themselves.

210-211 This is a typical Custom Chrome machine, built with parts from the C.C.I. catalog and decorated with the C.C.I. logo. The parts, signed "RevTech," come from a special series of performance elements and can be found on C.C.I. machines.

Drag Specialties

Drag Specialties, another specialized company, regularly presents its special machines during the big meetings at Daytona and Sturgis. The range of products displayed is developing gradually, and the same is true for the work carried out on motorcycles. In addition, the quest for originality prompts the experts such as Arlen Ness and Don Hotop, to indulge their imagination on an ever-wider series of bikes.

The specialist Don Hotop has already manufactured many parts and motorcycles for Drag Specialties, and over the years, these have been presented as exhibition machines in the company's catalog. They are generally manufactured on a Low Rider base; however, since Drag Specialties distributes parts for all types of Harley-Davidsons, other models of the brand are also customized and presented every year. The Softails bzcame more used and popular thanks to Drag Specialties. Thanks to the new ideas for transformations that have been unveiled and the production of new original parts for the Softail, its transformation has been made easier.

Touring machines, such as the Electra Glide, are particularly popular in the U.S., a country of vast open spaces where comfortable travel is particularly valued. Unfortunately, these motorcycles are too heavy and, above all, too high for small people. The Drag Specialties Research & Development department decided to take over the Electra Glide Classic case as well. The result was a motorcycle lowered to the minimum, using a tire several inches lower than on the original Electra Glide.

The modifications made to the frame to lower the motorcycle were based on techniques previously used on chopperized machines. For the front: a bigger caster angle and a Withe Bros kit allowed the fork to be lowered; and for the back, which was lowered to the same extent as the front: a reversed rocker arm, (the right arm was placed on the left with its head down and short dampers were adopted) – original solutions which reflect the expertise of the people who manufacture these very special motorcycles for Drag Specialties.

212 and 213 bottom ***Every year, Drag Specialties uses Low Riders and Dyna Glides in their catalog to demonstrate and promote new parts.***

212-213 and 213 top ***Softails offer a variety of options for transformation. The frame permits aesthetic creations of unquestionable quality.***

DRAG SPECIALTIES
DRAG SPECIALTIES
Drag Specialties

Harley-Davidson and the Aftermarket

From the middle of the '90s, Harley-Davidson became seriously involved in the marketing of parts and accessories for the motorcycles it manufactured. The department dealing with this sector was established within the company a long time ago, but this proactive approach is relatively new. In the time immediately after the brand and AMF had split up, exploiting this market was not the highest priority. At the beginning it was necessary to reinstate the range of motorcycles based on the Evolution engine, reassert control over and develop the dealerships, fight the illegal use of the Harley-Davidson name and logos, increase motorcycle production in order to meet the user demand, and so on, the result being that the parts and accessory market, although it was part of the Harley-Davidson Company's activities, were left to the competition. The parent company could not put up any resistance because it was unable to cope with all the problems at the same time; as a result, the aftermarket sector developed rapidly, to the advantage of companies such as Custom Chrome, Drag Specialties, Chrome Specialties, and others. From the beginning of the '90s, Harley-Davidson geared up, starting a successful clothing line which was eventually called Motorclothes. Then, in the middle of the '90s, a man named Clyde Fessler took charge of the Harley-Davidson accessories, grouping them as G.M.A. (General Motor Accessories), thereby linking his name inextricably to the growth of this sector.

During the '80s, Clyde Fessler had worked as sales manager for Harley-Davidson, then as the marketing manager. Subsequently he became the head of the "Rider Accessories Division," which then became "Motorclothes," and which he managed with a completely different approach. Success was just around the corner, and now he hopes for similar success in the accessory department. He began by studying the success of Harley-Davidson's competitors in this market and found out that not only do they do good work, but most of them concentrate solely on that sector and therefore are able to manufacture most of the parts much more rapidly than Harley-Davidson. Additionally, Harley-Davidson is bound by certain constraints. The company has committed itself to producing only parts which are legally fit for use on the road. This means abiding by regulations concerning noise, the size of headlights and indicators, etc. Fessler concluded, therefore, that Harley-Davidson should not compete directly with the independents, but should instead look to the future, where the company had a distinct advantage.

Only the company knows what new developments are in the pipeline; the G.M.A. department, as part of Harley-Davidson, can work on a range of accessories intended for the machines of the future, thereby having a ready-made range of accessories as soon as the new bikes are launched. Such accessories made their first appearance at the 1995 Sturgis meeting, where a new, "detachable" range of products – sissy bars, luggage racks, saddlebags, a

214 **_Great designers contribute to the quest for originality. Arlen Ness himself built this motorcycle for Drag Specialties; also the Harley-Davidson Motor Company worked together with Wyatt Fuller, founder of Razorback, to produce new parts._**

windshield that could be assembled and removed without tools in a few seconds, etc. – was presented. But there is much more room for growth as ranges develop around such sectors as Performance, Custom, and New models, and Fixed parts.

Harley-Davidson is aware that its parts are not as widely distributed as those of its competitors; there are about 4,500 stores specializing in products for Harley-Davidsons in the United States, but the number of official dealers, where G.M.A. products are sold, is only 600. In addition, G.M.A. put 280 new products on the market in 1996. By contrast, the number of new parts and accessories produced by two of Harley-Davidson's main competitors in this market in 1996 exceeds 2,800 – and there are about one hundred of these companies, of different sizes, manufacturing new parts for Harley-Davidson motorcycles.

On the up side, the Harley-Davidson Company has a strong hold in the franchising, gadget, and clothing sectors, as the brand benefits from a worldwide reputation.

215 top **_Electra Glides, for a while overlooked by designers, have increasingly benefited from a specialist's touch, as with this motorbike by Drag Specialties, customized beyond the usual chrome plating and luminous arrays._**

215 bottom **_Sometimes, for special meetings, Harley-Davidson will recall its glorious past by positioning an old machine in a period setting or working one into a party, as here for the celebration of the eightieth anniversary of the magazine_ The Enthusiast.**

***216-217 top** A 1992 Harley-Davidson FLHS, entirely revised and corrected by Downtown Harley-Davidson of Seattle, Washington. All the bodywork was shaped by hand.*

***216 bottom** This machine, with distinctly radical bodywork, is based on an FXST in principle and conception. Its fork prolongs the line of its engine block, which rises up to cover the frame.*

***216-217 bottom** Certain technical solutions proved to be original, such as the rear shock-absorbing system, aided by a torsion bar. But it is the personality of this motorcycle that is particularly impressive; the unusually voluminous exhaust outlet highlights an aggressive animal look.*

Customization in Seattle

These two motorcycles opened up new possibilities for customizing and transforming Harley-Davidsons, as they set themselves free from the limitations established by the parts made available by great designers.

Russ Tom, from Downtown Harley-Davidson, a dealership located in Seattle, Washington, decided to create two totally different motorcycles. It is worth pointing out that he was a long-time friend of Arlen Ness and that he had literally swooned after seeing Arlen's Nesstalgia, a motorcycle that convinced Tom to go even further with motorcycle customization.

Working with an FLHS – or at least on what was left of it after its complete disassembly – a new frame was created that was over three inches longer and increased the caster angle by about six degrees.

Weeks were spent hammering and shaping the mudguards and tank. At the end of the work, it was realized that the rear mudguard was a bit too wide so, for this very simple reason, the feature was extended even more in order to house the rear exhaust outlet.

The motorcycle tank was lengthened and modified as Terry, an aluminum worker at Downtown Harley-Davidson, had spent time shaping the mudguard and tank. He was open to all possibilities and did not hold back, pushing his creation even further. Air-scoop style body work underlined the exhaust pipes on both sides of the motorcycle, which ended in two huge megaphones, the biggest ever seen on a Harley-Davidson. A small fork head and a profiled mudguard gave a look redolent of Japanese motorcycles, an effect that is promptly forgotten as soon as the machine is considered as a whole; its aesthetic is unique.

***218 left** Sbarro hubless wheels and steering linked into a hydraulic system are solutions which may be used on production motorcycles in the near future.*

***218** To obtain the expected aesthetic result, even the engine is fitted out with elements to make it look as if it comes from the future.*

***218-219** The futuristic lines of this prototype mirror the motorcycle's state-of-the-art technical solutions as well.*

Apache Warrior

The Apache Warrior, presented during the 1995 Bike Week at Daytona Beach, was a revolutionary machine, conceived around a Harley-Davidson engine block. Styled as a state-of-the-art motorcycle, it was manufactured by Next World Design and was fuel injected to obtain 110 horsepower from the big V-Twin which they had carefully fitted out. Other advanced techniques used here included Sbarro wheels (without hubs) and power steering, with the whole machine weighing 570 pounds. This is a motorcycle which has effectively passed to the production phase, as more than twenty were ordered as soon as it appeared in public.

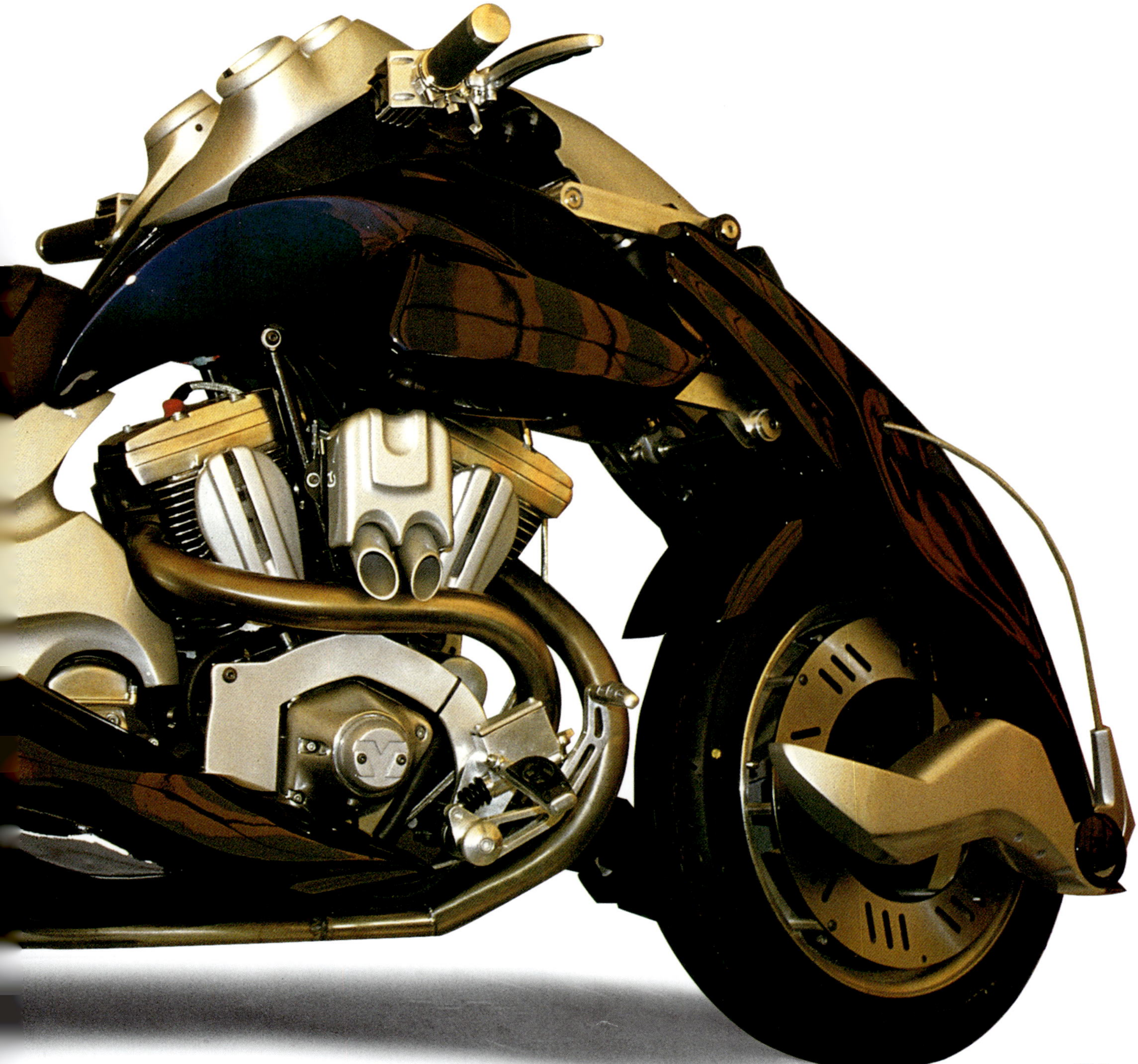

Arlen Ness: Master of Motorcycle Design

Arlen Ness: more than just a name, he is a guru for the fans of beautiful motorcycles. His machines, carrying the Harley-Davidson label, are sculpted, evolve, and become real works of art in the hands of this international design specialist.

Arlen Ness has been a professional creator and manufacturer of exceptional motorcycles since the beginning of the '70s; he created a new style of chopper and modified motorcycle that gradually prevailed among many other creators, and eventually was referred to as the "Ness Style" – a considerable success for the young truck driver who began this paint and transformation work on Harley-Davidson models as a hobby at the end of the '60s. But this restless spirit was not content with simply manufacturing identical motorcycles to meet customer demand. He continued to question his work, pushing the limits of feasibility further and further.

Ness quickly moved away from the "Frisco Style" shaping the development of choppers to create his own style of low, lengthened, and slender motorcycles, which irresistibly conjured up the Dragsters. This is how the "digger style" of the '70s took shape.

In 1979, his creativity reached a frenzied pitch when he coupled two Sportster engines, thus creating a 2000 cc motorcycle with a semi-supporting engine, with a positive displacement blower and two double-cases grazing the ground. The suspension was based on torsion bars and the steering was from a car. "Two Bad" won all the shows it took part in. Throughout the '80s, Ness monopolized the trophies in all the motorcycle shows, always surprising his fans with new and ever more daring transformations; he pursued his quest for originality, fitting a turbocharger on one machine, producing another with a "retro" look, creating a futuristic motorcycle on a third – and, of course, he always travelled behind the wheel of his newest creation. The Low Riders replaced the Sportsters little by little as Arlen Ness's favorite bases for transformations. The FXR Police models, fitted with Evolution engines, rapidly became his forte during the second half of the '80s.

In this period, through meetings and motorcycle shows, he discovered Europe. He appreciated the continent, and, as soon as he could spare some time, he returned. At the beginning of the '90s, he revisited the countries to which he had traveled for some shows, such as Germany, Great Britain, and Sweden, but he also discovered Italy and France, two countries he particularly liked. He went back for a journey on motorcycle over the Alps, then visited the Pyrenees, the Languedoc region, and Spain, and took part in a trip around Corsica in 1996.

As Arlen Ness's company started to grow,

Arlen's son, Cory Ness, followed his father's example more and more closely. Arlen consequently had more free time both for travelling and for creating motorcycles with increasingly original designs. A lover of beautiful cars, his designs often paid homage to exceptional models. In 1990, he unveiled his "Nessrossa," a red and chrome-plated 2000 cc V-Twin motorcycle, with rear side covers replicating the design of the Ferrari

***220 top** Arlen Ness presents one of his fabulous machines, this one fitted out with bodywork that he designed to pay homage to the Bugatti.*

***220 center and 221 top** The Ness Convertible Harley-Davidson has two facets, one of the touring motorcycle type with enveloping mudguards and a big tank, the other is a performance Street Racer with mudguards and other bodywork removed, and has a lighter line.*

***220-221** This classic Ness creation reuses elements well-known in his production line.*

***221 bottom** There are two examples of "Mona Lisa," with the rear wheel entirely unveiled when viewed from the back. This design appealed to a Spanish Harleyist, who ordered an exact replica of Arlen Ness's original creation.*

Testarossa. Various other vehicles followed, based on the concept of "dressed" motorcycles, with or without detachable body, which grew closer to the idea of the car body. At the end of 1994, he put this concept definitively into practice, by presenting his "Nesstalgia," a Harley-Davidson evoking a '57 Chevrolet. Then in January 1996, he unveiled his incredible "Bugatti-Ness," which epitomized the concept, both in terms of aesthetics and quality of designs, and allowed Ness to pay tribute to two of the machines he loves the most.

Today, thanks to his work and his broad-mindedness, Arlen Ness is even accepted by the Harley-Davidson Motor Company, which considers him an ingenious designer, a positive force for Harley-Davidson, and not a competitor. Arlen Ness, whose many transformations based on Harley-Davidson motorcycles are already displayed in American museums, now has a worldwide reputation, thanks especially to his recent spectacular success in Japan. After first being recognized in California, then in the whole United States, before spreading over the North American continent and creating unprecedented excitement, both in Europe and in Indonesia,

***222 bottom** FXR bases like this one, of the Low Rider type, often come from Sport Glide motorcycles once owned by police forces, who sold them after renewing the stock. In these cases, various frame elements have often been modified.*

***222-223** These classic lines correspond to those of motorcycles built regularly in Arlen Ness's shop, where he adds touches designed to meet the requirements of customers, friends, and aficionadoes, all of whom want to ride a motorcycle signed by Arlen Ness.*

223 center ***The line of low, long machines, originally based on the FXR frames of old Low Riders, is now based on Dyna Glides.***

223 bottom ***This machine, based on an old Sportster engine, perfectly corresponds to the exceptional creations that Arlen Ness has made popular, building his reputation as an extraordinary designer.***

Arlen Ness has been fascinating the Japanese with his products for Harley-Davidsons for more than a year.

Interestingly, Japan, itself the top motorcycle manufacturer in the world, is conspicuous for its large consumption of Harley-Davidsons. Arlen Ness has estimated that it will be his principal export market before 2000. Arlen Ness, a man of exceptional talent and character, who continually renews his ideas and puts into practice the inspiration of the day before, deserves all the success he has received. His most recent forays into the far Eastern market must be a thrilling development for the man who could only receive customers at his first small shop in the evening, because during the day, he still had to work as a truck driver to support his family.

224-225 Engine modifications sometimes affect performance, but primarily they are undertaken for the sake of looks. Compare here an original Big Twin Evolution engine with the improved engine, provided by Arlen Ness.

225 top ***Arlen Ness's son, Cory Ness, works for his father's company and also styles motorcycles. Here is Cory Ness's "Wedge," which anticipated future bodywork from Arlen Ness himself.***

225 bottom ***Machines coming out of Arlen Ness's shop prove to be easy to date, since he signs their engine parts. Here, the air filter and the rocker arm push rod covers are the most visible Ness elements. On a more recent machine, the gearbox and ignition covers and the cylinder head covers would signify the Arlen Ness touch.***

226 This Sportster engine, fixed on a rigid frame, has lines that evoke the look of old Harley-Davidsons. It is an original creation utilizing a Springer fork, a tank design of the 1920s, and a handlebar that sweeps back. The whole thing has then been painted bright red, to turn the machine from a fake old bike into an ageless motorcycle.

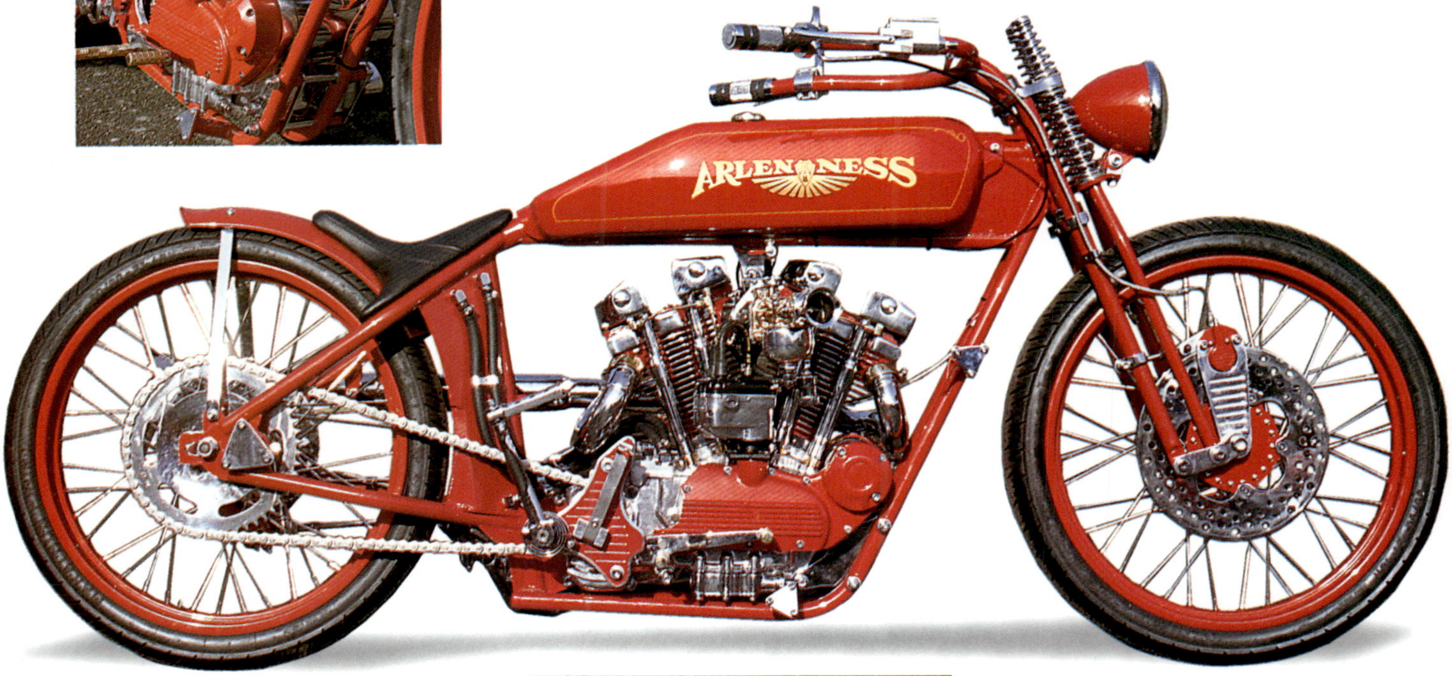

227 The Sportster engine is often modified, both for looks and for performance. The gearbox housing has been cut so that only a plate protects the pinion, which still supports the kick-start. The oil sump acts as the rear frame support and the extension of the rear mud guard, while the braking system is based on hydraulic disks and calipers, discreetly hidden in the motorcycle body.

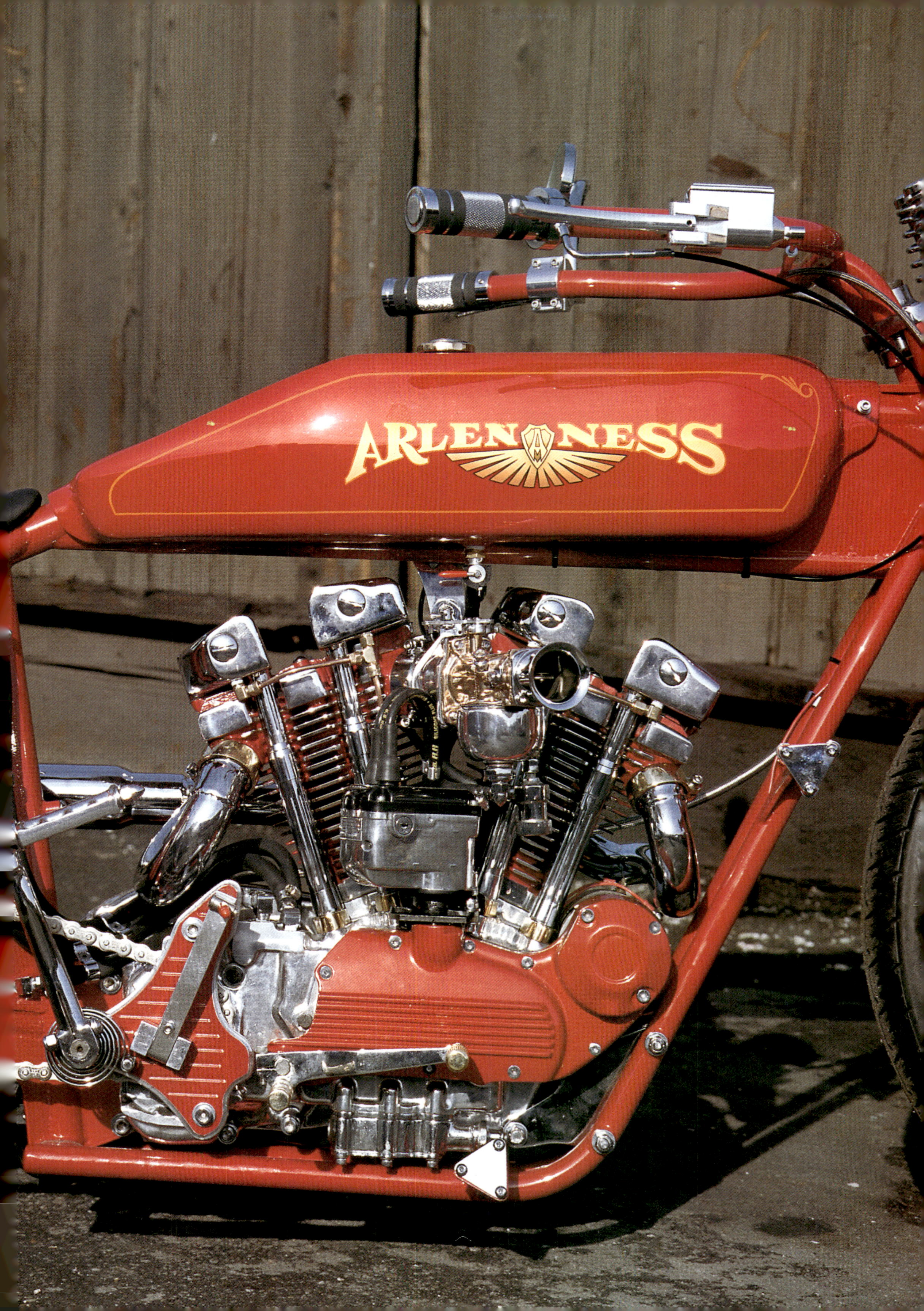
ARLEN NESS

228-229 The "Ness-Stalgia" combines two mechanical legends. It was built by Arlen Ness in cooperation with Carl Brouhard and designed to evoke the look of a 1957 Chevrolet Bel Air. The back of the motorcycle looks like the back of the Chevrolet, while the front part mirrors one of the car's mud guards.

228 bottom The Pepsi-Cola motorcycle: this model was customized by Arlen Ness for a friend who works with Pepsi and owns many Ness-built motorcycles.

229 bottom The "Nessrossa," an amazing machine fitted with a huge V-Twin 2-liter engine, four Dell'Orto double carburetors, two Jerry Magnuson Blowers NA180's, and a nitrous oxide bottle on each side. Arlen Ness himself would sometimes ride this machine between shows.

Bob Dron

230 top *The third Royale, built by Bob Dron, is much lighter in line. With its whole bodywork finished in colorless metallic paint, it proves to the most skeptical that it is not made of plastic.*

230-231 *Royale Number 2 has the same mechanical base but a more enveloping body, whose rear sides include saddle bags. Made entirely of metal, it is fitted with double mud guards and a double-lens headlamp. It even sports saddlebags and inlays of gnu leather.*

231 top and bottom *The first Harley-Davidson Royale, built by Bob Dron, was a Softail fitted out with a fretworked body and a saddle made of ostrich skin.*

231 center *Bob Dron and his wife, Tracey, are the official Harley-Davidson dealers in Oakland, California.*

Whether on two wheels or four, Bob Dron knows how to attract attention! After working in the car field and then branching out into choppers and custom manufacturing, he has become one of the most important dealers in the world.

Faithful to the brand since 1981, he has, on five different occasions, been awarded trophies by the company, among them, in 1991, the highest Harley-Davidson decoration: the Bar and Shield with special mention.

Since the release of his red and chrome-plated Heritage Royale, fitted with aluminum body work evoking the fineries of the elegant European car brands of the '30s, Bob Dron has been considered an unquestionable world master in the art of customization. The heritage Royale was followed by the Heritage II model, fitted out with aluminum body work and painted in royal purple, in 1994.

These motorcycles are the result of the combination of Bob Dron's two passions: the customization of cars and Hot Rods and that of motorcycles. This attraction for original cars can be traced back to the very beginning of the '60s, when he used to polish cars and boats for exhibition just before their admission to the famous Oakland Roadster Show. In the '70s, this passion developed into a profession when, after leaving his home town, he founded the American Choppers Enterprise (A.C.E.), a company which specialized in designing and manufacturing custom motorcycles, show bikes, street choppers, and sidecars. Bob became gloriously successful in his field and one of his machines was even on the cover of the first issue of *Chopper Magazine*. He met Arlen Ness, who opened a new view of the chopper world to him, as Ness had just created the Californian "digger look". Bob immediately realized that Arlen was both diligent and passionate, and in Bob's view, he is a master of innovation.

The Royal Heritage 3 is based on the same

principles as the previous versions, but with its "body art" pushed even further. Its forms are more profiled and less massive, and highlight the mechanical parts by further reveiling them. The aluminum forms are further modified and state-of-the-art technology is present in terms of the engine development and of the frame, which uses a Buell reversed fork. Nonetheless, this kind of body work is usually the product of automobile professionals rather than motorcycle professionals and this customization has proved to be closer, in aesthetic terms, to conventional machines than the previous Royale models.

***232-233* Cyril Huze created this "Road Star," a modification of the Road King, for everyday travel. It is a motorcycle which is good-looking, practical, and full of personality.**

Cyril Huze

Cyril Huze is a Frenchman who moved to America more than ten years ago to live out his American dream.

Cyril: "There was nothing original in a French teenager at the end of the '50s dreaming of America with all its stereotypes: rock 'n' roll, cowboys, the open spaces, jeans, the films, the hot rods, and obviously the Harley-Davidsons running along Main Street. Similarly, there was nothing original in becoming one of James Dean's fans. He was young enough to be close to all the world's teenagers . . . and old enough when he died to become a hero.

By contrast, the fact that I could not content myself with simply watching the three cult films *East of Eden, Rebel Without a Cause,* and *Giant* was rather original. Like all young people, I was happy believing that the characters interpreted by Dean were not simply film roles. Thus, even before examining his life, I was convinced that this angel-faced guy, who behaved like a rebel in the films, was also a rebel in real life. But I also wanted to know why.

My youth was difficult and agitated. Was this also the case with his?

The hardest thing was to find out information because all the interesting things which might concern him were written in English – a language I could not yet speak. As Christmas was approaching, I explained my problem to my grandmother. She understood the problem perfectly and gave me two

***232 bottom* Reminiscent of James Dean, the "Deanager" was painted by Chris Cruz, an exceptionally talented artist from Florida. Cruz's work focused on a thorough finishing, an oscillating arm, and some engraving work in a modern design, repeated on all the metal parts.**

autobiographies by James Dean with . . . an English-French dictionary. So I started learning English at the age of eleven, through James Dean's life. Thank you, Jimmy."

Cyril emigrated to the United States in 1986; years of work had finally allowed him to make many of his American dreams come true. One of these dreams was that of building a Show-Bike dedicated to James Dean and, logically, its name "Deanager" gave it a nostalgic note, its blue was that of Jimmy's eyes and its tires had white walls.

"Deanager" was built from an Evolution engine, but Cyril opted for a Shovelhead block for his other exceptional bike, "Miami Nice," which was fitted with an exhaust pipe with closed ends. (Cyril had the pipes cut at the sides so that the gas could escape from there.) The dampers were also quite original: they were hidden, but still existed and fulfilled their function perfectly. These elements, together with the saddlebags and many other details, were his brainchild. Thus, he was among the first to conceive the growing trend in customization, in which the work is shared among the designer, the manufacturer, and the machine assembler.

Cyril considers himself a designer, and he maintains that, at the present time, there is a sort of rivalry between those who assemble customized motorcycles and those who design them. Up until recently, mechanics used to assemble motorcycles without following real plans; today, designers design machines from start to finish and submit the related plans to the professional mechanics. Now, there is a shortage of designers on the market, and,

***233** Although this "Deanager" Softail is not the first creation by Cyril Huze, this motorcycle brought him public attention and professional recognition as an exceptional designer.*

in their rivalry with "bolt assemblers," they are likely to win, because they always seek new inspiration.

According to Cyril, a Harley must remain a Harley, and all the motorcycles he designs have the Harley-Davidson look. They are the expression of his personality; he transfers his emotions to them, gives them a name and, from there, he decides which colors are suitable to characterize the chosen themes. The aesthetic is then drawn from the chosen theme and name, and underlines and emphasizes the form, greatly influencing the final result. The engraving, too, reflects the theme. The theme for the motorcycle "Miami Nice" is Art Deco, a style that was invented by the French and modified by the Americans, and is so prevalent in Miami.

234-235 "Miami Nice," a 1997 FL Electra Glide with a Shovelhead engine, evokes the French Art Deco fashion that invaded Miami during the "Belle Epoque." Its look is very sophisticated, with swept-back saddlebags, while its construction incorporates state-of-the-art technique. The rear dampers slide inside chrome-plated tubes, entirely disguising them. The exhaust pipes, closed at their ends, release fumes through invisible rear side slits. Every bit of chrome-plated metal on this machine is engraved.

Corbin Bodies

Ron Simms, from Bay Area Custom Cycles, produced the first bike based on body work elements which had been created by Mike Corbin. Corbin worked regularly with designers and transformation specialists, who focused on the particularly interesting possibilities offered by the Corbin body to those working in the customization field. Thus the "Warlady," whose design is based on the "Warbird" kit by Corbin, is a motorcycle with performance worthy of road and even sport qualifications.

In building the Warlady, Ron Simms was allowed total freedom – except for modifying the Warbird kit, for which the machine was supposed to act as a promotion tool. It is worth remembering that the purpose of the Corbin monocoque kit was primarily to allow anybody to transform the look of his motorcycle by bolting the parts together, with no need to change anything on the machine.

Ron Simms chose an aviation theme, in particular to a World War II fighter plane. The fairing was decorated with a pin-up in the style of Vargas, the famous *Playboy* magazine illustrator whose drawings had decorated all young American males' bedrooms and vehicles for decades. To further emphasize the aviation look, the upper part of the motorcycle, which was visible from above, had desert-camouflage paintwork, while the bottom part was pale blue so that the vehicle would blend with the sky during its flights.

After this creation, many Harley-Davidsons were decorated with Corbin Warbird kits, which were often combined with saddlebags also designed by Corbin; the motorcycle personalized in this way by Arlin Fatland from Two Wheelers proved undeniably successful.

***236 top** Warbird bodies give a performance look to Big Twins and Harley-Davidson Sportsters.*

***236 center** Corbin's "Warlady," based on his "Warbird" body, is a creation by Ron Simms aimed at promoting this new design for Harley-Davidson.*

***236 bottom** By completing Corbin's "Warbird" body with polyester saddlebags, a Harley-Davidson can be fitted out in an original way. The result proves to be quite successful, particularly if the machine is customized by a professional like Arlin Fatland.*

Cycle Savage Works

Cycle Savage Works has evolved, improved, and diversified its work thanks to Lonnie Cantrell and his partner Anthony Verdibello, who form a top customization team in Florida. Lonnie started working in a small windowless room on Pompano Beach that was originally a goods store which had been turned into a workshop. Now he is employed in a real store and, above all, he has enough room to meet his customers' growing demands. Although most of his customers are based in Florida, orders are coming in from further afield, thanks to the renown gradually achieved by Cycle Savage Works throughout the United States.

***237 top** All types of Harley-Davidsons benefit from some work, such as this Dyna, whose final look only vaguely resembles its original appearance.*

***237 center** The classic reconstruction performed on this Softail was requested by a German customer who rode the motorcycle in Florida before taking it home to Europe.*

Lonnie is used to modifying all the original parts of the motorcycle he works on. This white machine is a Dyna Glide whose components have all been modified, reworked, or replaced. The wrapping mudguards have been obtained by assembling and modifying two Dyna Glide rear mudguards for the front and by doing the same thing for the rear – a total of four mudguards to create two unique models.

The dampers, placed well back, are of chrome-plated metal; they slide inside metal tubes, even though they retain total freedom of operation. These tubes were originally exhaust pipes and have been capped with the old FLH damper covers. It's a technique that required extensive research and many tests to perfect, but now it is possible for the machine to run as reliably as an original motorcycle.

***237 bottom** Cycle Savage Works, a fairly recent Harley-Davidson customization company, has grown rapidly, producing models based on a variety of themes, including this personalized Softail.*

Donnie Smith

Donnie Smith is one of the most renowned bike builders in America, but he is less popular abroad than most of his counterparts. This may be due to the fact that he takes the time to personally build the machines; he does not try to produce huge quantities. As a result, he manages to maintain a constant quality level in his products.

This approach stems from his wealth of experience and his love for metal and good work. Always in search of new custom fields to develop, he has explored all the present trends, leaving his personal mark each time.

Donnie Smith was born in a small town in Minnesota and later moved to Minneapolis. He was educated at a farming school, but as a youth spent most of his free time with his friends, exploring their mutual love of metal working.

Subsequently, he worked for Honeywell and took an interest in drag racing, and spent every weekend on the tracks with Willys Gasser's team. During the week, Donnie and his teammates manufactured parts for other drivers, and this rapidly turned into a full-time occupation. Soon they decided to open "Smith Bros. and Fetrow," a store opened with the purpose of financing a dragster while they built it. All the work was focused on dragster cars – until the day they were asked for motorcycle parts. The parts were for some friends, but, as demand grew, the related accessory activity rapidly became the main occupation for the store, which in the end, turned into a workshop specializing in repairing and manufacturing motorcycles.

"Smith Bros. and Fetrow," had become a chopper specialist and was among the most famous for the quality of its work. Donnie manufactured numerous original parts, such as frames and forks, until the day the demand for chopper parts calmed down, at which time Donnie and his friends closed the store and decided to go their different ways. Donnie took the opportunity to travel and even thought about retraining; but he loved that type of work too much, and soon encouraged by the level of demand from customers who wanted to have a

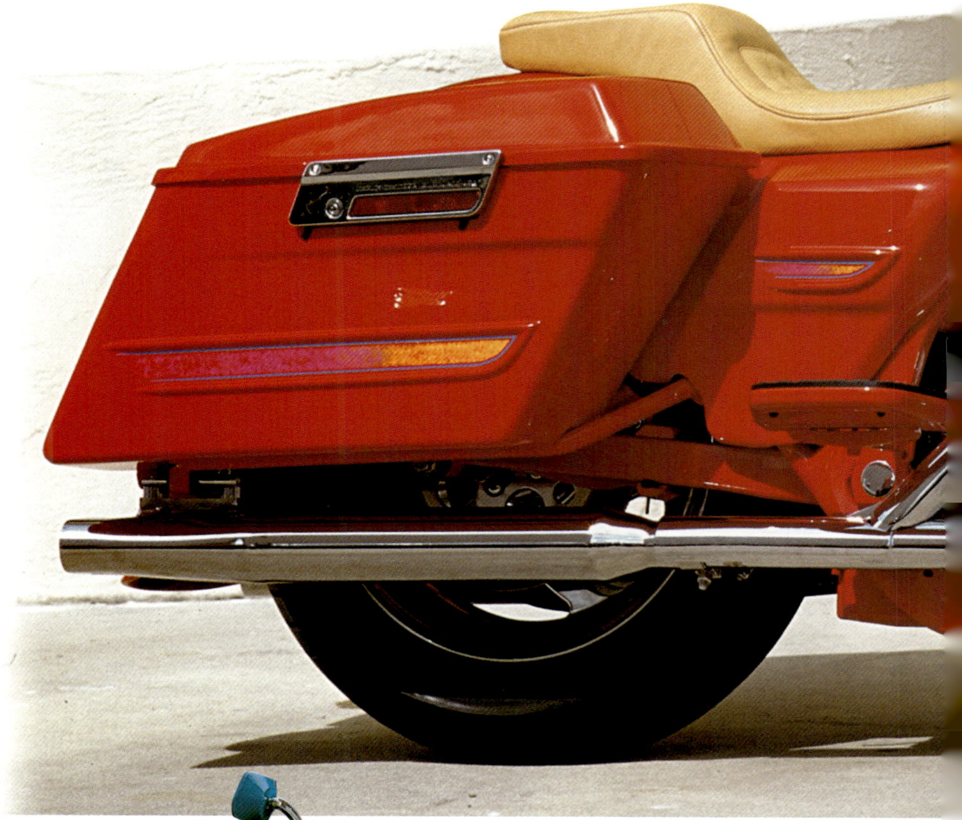

***238 top* Donnie Smith poses on his Road King during Bike Week at Daytona Beach, enjoying an overdose of sun, far from Minnesota's winter snows.**

***238-239* The Road King models appeal to an increasing number of designers. They offer practicality, with their saddlebags and removable wind-screen, but they also offer improved lines, filling in the empty space between the saddlebags and the rear mud guard.**

machine built especially by him, he settled down by himself in a new shop. In this new workshop, he not only manufactures custom motorcycles, but he takes his time to discover new facets of the field; he even loves explaining the modifications he carries out, the procedure he applies, or why it is necessary to reinforce a certain part at a particular point. He is a real technician who knows his field inside and out, which explains his present success and international recognition he has garnered as a master of the subject in the manner of Arlen Ness and Dave Perewitz, just to mention the best known of the Hamsters.

***238 bottom* Before developing an interest in the Road King, Donnie Smith applied his skills to the Electra Glide and other Harley-Davidson models.**

***239 bottom* Donnie Smith is most famous for this motorcycle, which he derived from an FLHT. Its thin wheel nuts and the handlebar with a speedometer, are particularly popular.**

Hardly Civilized

Every year, at the great American meetings, and in particular at the Daytona Bike Week, the bike shows act as springboards for launching fantastic new creations. Here, both amateur and professional designers unveil their masterpieces, which may be finished or may still be works in progress.

In 1995, two men named Eddie and Simon, who had formed a partnership called "Hardly Civilized" Incorporated, stood out from the crowd when they exhibited two highly original Harley-Davidsons, on which their owners travelled everyday.

These machines had been assembled in Liberty, South Carolina and had been fitted with a unique exhaust system. This impressive innovation was built from a single pipe connecting the two

240 Hardly Civilized immediately stood out from other customization specialists, offering particularly original exhaust systems on Harley-Davidsons.

cylinders, without a baffle, and cut sideways in the middle. Contrary to expectation, the noise is kept within acceptable limits, and more importantly, the engine works perfectly and performances have been increased.

But these are other innovations as well. Hardly Civilized focuses on each specific motorcycle they transform. They rarely use parts which are widely available on the market, and when they do, they transform and personalize the parts before using them in their creations. One example of this is a handlebar which is fixed under the tee, thus obviating the need for a linking bar between the arms.

Since their debut in 1995, Hardly Civilized have presented their innovations and their creations at every big American meeting, consistently pushing the limits of design.

240-241 This exhaust system connects the two cylinder outlets to obtain a deep but moderate sound and improved performance, according to designers.

241 top left Few customized machines and even fewer road motorcycles have adopted this type of exhaust. Nevertheless, its aesthetics draw attention at shows.

241 top right A Harley-Davidson decorated to match the label design of Jack Daniels – definitely a "Biker" look.

242 top and 243 bottom Using many custom-manufactured elements, Ron Simms of Bay Area Custom Cycles builds Softail Harley-Davidsons, mounting huge rear tires and preserving a belt-type secondary transmission.

242 bottom The "Fandango," with which Ron Simms won the Oakland Show in 1993, was initially a Softail. A radically different motorcycle from Simms's other creations, this cycle took part in many subsequent events.

Bay Area Custom Cycles

Ron Simms's history as a designer began a long time ago, when he purchased a dismantled Sportster belonging to one of his neighbors. He wanted to reassemble that machine for his own use, adding a few personal modifications, which then multiplied as the work went forward. When the motorcycle was finished, he was offered a much higher sum of money than what he had paid. This gave Ron the idea of doing it as more than a hobby. He constructed several machines in his own garage and later, in the '70s, decided to open Bay Area Custom Cycles. The shop he found in Hayward, California had originally been built in 1956 for a Harley-Davidson dealer, who was now just closing his doors. Ron Simms moved in and has never left the place. He carries out work of acknowledged quality, endowed with personal style. It's a look that he alone can provide and which has fascinated an increasing number of people, including stars such as Neil Young, Greg Allman, and Boz Scaggs. His approach to motorcycles is focused on both the aesthetic and the technical.

This approach dates back to the time when Ron Simms opened his shop, and decided to create a look that was different from that of the local choppers. He created a customization aesthetic which was entirely new for the time. He built motorcycles with a different steering stem angle and a shortened fork, thus implementing his first low and imposing motorcycles. The real B.A.C.C. style was refined with the passing years; and, at the beginning of the '80s, Ron Simms increasingly engaged himself in the manufacture of the elements used in composing his motorcycles. Since then, he has kept the same policy; and when the parts do not come from his workshop, he always selects models which he thinks are produced by the best manufacturers so as to produce motorcycles of unquestionable quality.

242-243 Traditional Harley-Davidsons – Low Riders or Dynas, with Evolutions, Shovelheads, or even older engines – easily adopt the lines of good-looking motorcycles mainly intended for riding.

Rick Doss

Rick Doss, the Harley-Davidson parts specialist based mostly on the East Coast for C.C.I., regularly produces machines for Custom Chrome in addition to his personal parts. Doss deserves to be recognized because he is living evidence that anybody with a bit of taste and, above all, ideas can impose himself on the market of transformed Harley-Davidsons.

Rick, who designs and manufactures his own parts to transform and customize motorcycles, is also, like John Reed who works in the Custom Chrome research and development department, a designer-manufacturer of new products for Morgan Hill. Every motorcycle that Rick Doss builds, using parts which he designs specifically for that machine, allows him to test the market during big shows like Daytona and Sturgis. The bike is exhibited, and depending on the reactions of Harley enthusiasts, its parts may be launched into production. Thus, the motorcycles act as showcases for the new parts which will to be available in the C.C.I. catalog.

Initially, Rick owned a service and repair workshop for cars and trucks. However, being very fond of motorcycles, every time he bought a Harley he could not help modifying it, first changing the color and then some of its parts, such as the tank. At first, his work was mostly carried out at home in his leisure time; then it gradually encroached on his working hours, because he was always looking for new ideas that went against the various standard concepts of the time. (For instance, when everybody wanted chrome plating and polished parts, he painted them black.) Then, more than twenty years ago, he decided to sell his car and truck workshop and settle down in a new place to work on Harley-Davidsons. Today, he is still in the same place and still works in the same way, acting on impulse according to what he feels like doing with the motorcycle he is about to build. He is capable of spending hours on a component and then abandoning it when he realizes that its final line is not suitable for that particular motorcycle. He always aims to design machines which are both good to look at and easy to ride. His greatest pleasure is when a Harley enthusiast admires one of his motorcycles and imagines himself at its wheel.

244 The Softail has been the Harley-Davidson most widely customized, following many themes and interpreted by many designers.

245 Rick Doss, a renowned designer who creates parts for Custom Chrome Incorporated, shows the motorcycles he builds in a variety of American shows. They can also be found on the pages of C.C.I.'s catalog.

CHROME

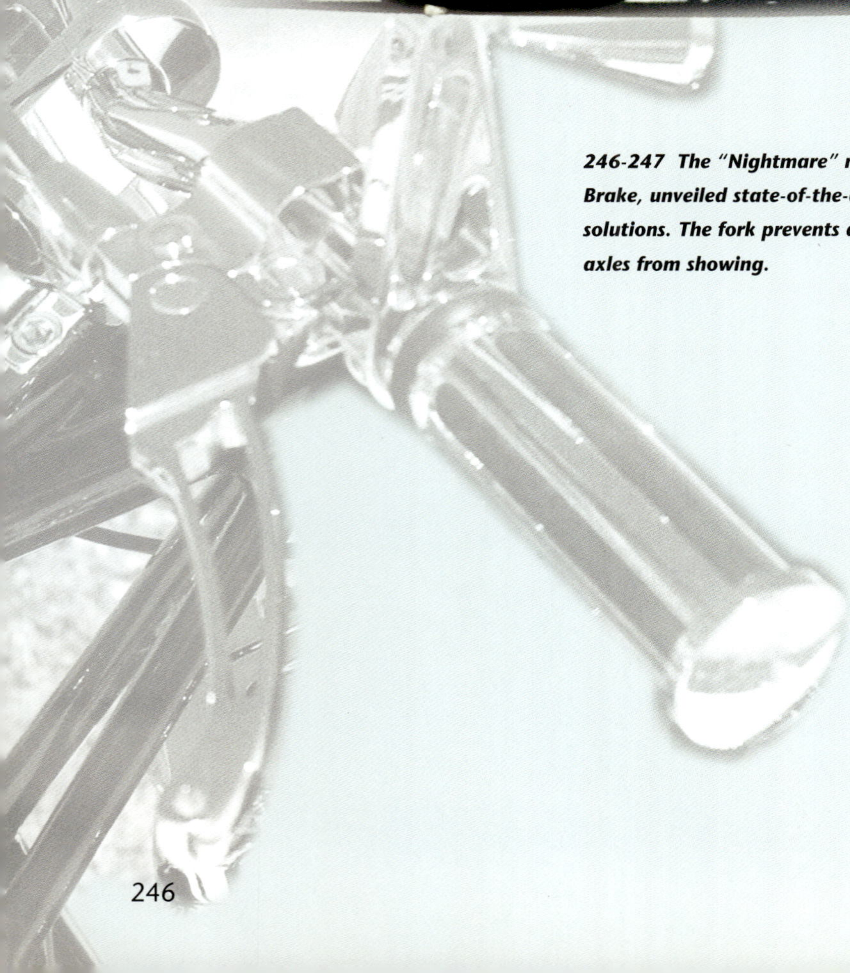

246-247 The "Nightmare" model, by Jay Brake, unveiled state-of-the-art technical solutions. The fork prevents all bolts and axles from showing.

Jay Brake is a famous company specializing in braking systems and controls which are placed well forward. It likes to unveil its exceptional and thoroughly researched motorcycles, such as the famous "Originator" and "Nightmare," during the big American shows.

The Originator is a motorcycle designed by Jay Brake, whose brake calipers do not act on the central brake disks but directly on the rims and whose fork hides the front wheel axle completely.

The Nightmare adopts the same type of fork, a very similar design for the frame, and practically the same front. But the back of the motorcycle is much more tapered and integrated into the general line of the motorcycle, whereas the Originator has a more truncated back line.

The name "Originator" was derived from the new braking system adopted, whereas "Nightmare" is related to the huge amount of work that the design and implementation of this motorcycle required and, above all, to the need for finding solutions to problems which at first appeared impossible to untangle – as is usually the case when an exceptional motorcycle is being built, maintains Jay Brainard, the owner and eminence grise of Jay Brake Enterprise.

Motorcycles have always been a passion for him and working in a field related to Harley-Davidsons is an ideal opportunity for manufacturing fantastic motorcycles.

***247 bottom** Jay Brake's first exceptional creation was this "Originator," equipped with brake calipers, clamping a disk on the perimeter of the rim.*

J.P. Poland has become well-known in the field of transformation by manufacturing a Harley-Davidson called "El Tigre," which was the product of his encounter with Dan Meyer, a styling specialist and one of the designers of the Absolut vodka bottle. Dan Meyer understands the requirements of creating a design which will appeal to a large audience, and thus, beyond castings reinforced by the addition of metal in certain points of the mudguards, he suggested tiger-striped paintwork. It was very successful. After being displayed in the window of a cafe in Florida, this motorcycle appeared on the *Regis and Kathie Lee Show*, after which it was invited to the Harley-Davidson Café inauguration in New York City. As a result of the success of the first El Tigre, the Café ordered a similar machine on a Sportster base.

Then J.P. decided to undertake a new project, which would be based on the same principle but would focus on another theme. He and Dan Meyer purchased a new FLHTC 94 and started transforming it into a dragon. As they had more time now, they were even more painstaking and built a motorcycle which was trimmed and decorated in every detail according to the dragon theme. The bike was fitted out with polyester and resin and the casting, assembly, and molding work, as well as the aluminum parts, were by J.P. and Dan, who also carried out the tapping and setup, with a final result of spectacular aesthetic effect.

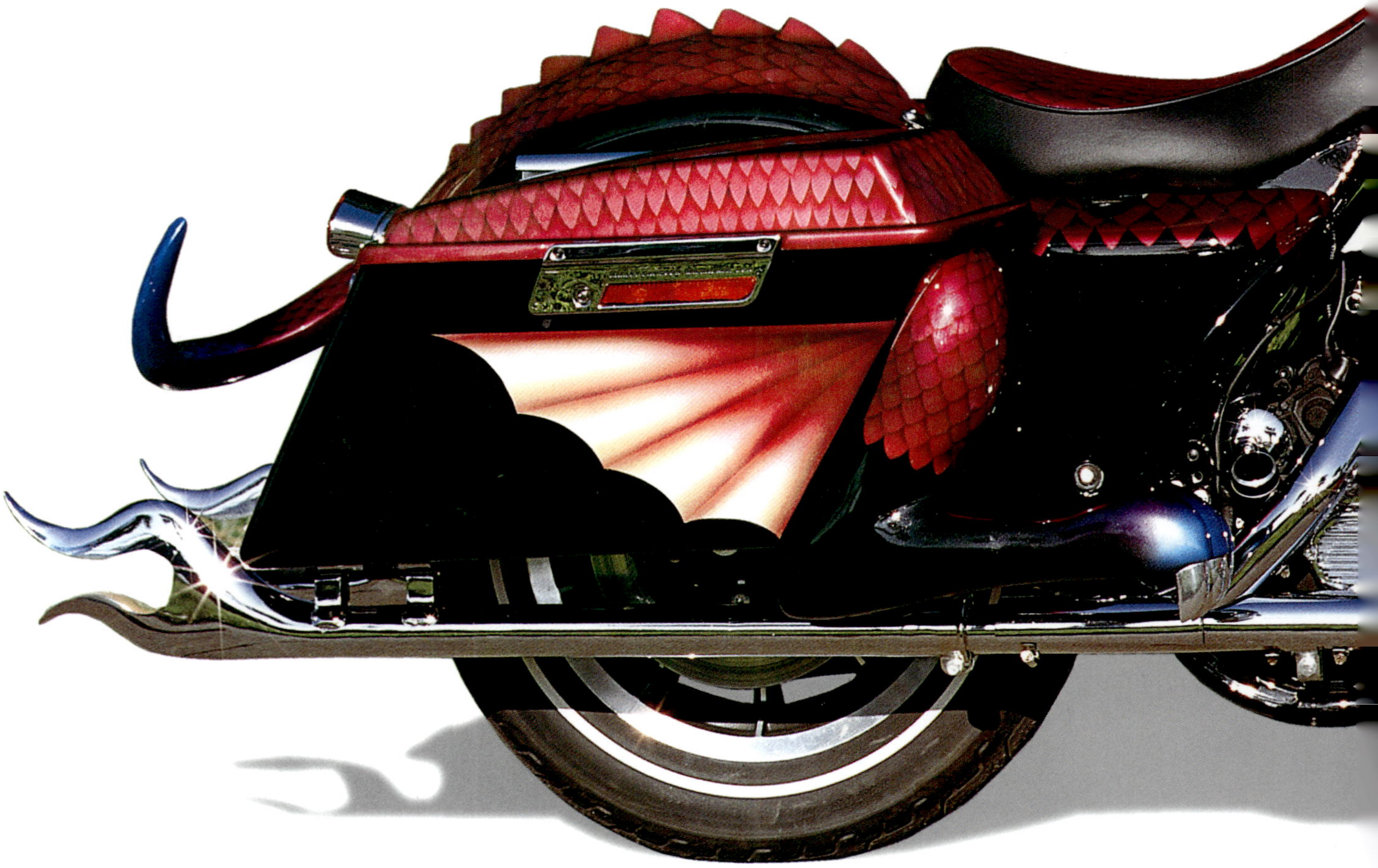

248-249 The "Dragon" motorcycle is a 1994 FLHTC with bodywork in polyester and resin. This motorcycle could be ridden, but J.P. Poland decided to reserve it for shows and promotional display.

DREAM THEMES
407·684·8155

SEP FLORIDA 93
942740

250-251 J.P. Poland first built "El Tigre" and presented it during Bike Week at Daytona Beach. Built on a FL base fitted with a Shovelhead engine, the body is composed of reinforced polyester elements. Later, Poland built a machine along the same theme, using a Sportster base, for New York's Harley-Davidson Café.

252-253 An entirely modified 1994 FXR: the body consists of a single piece and the engine look has been modified, especially in the size of its cooling fins.

252 bottom Here, Bob Lowe used a Jay Brake fork, disguising both axle and bolts.

Bob Lowe first entered the custom motorcycle field with a 1993 Fat Boy. This first motorcycle was built when Bob Lowe already had several customized cars; he used it to extrapolate the styles existing in the automotive sector and adapt them to motorcycles. After the success of this first machine, which was rewarded with prizes during various shows, he undertook the creation of a second motorcycle on an FXR base, as few people had done such extensive work on this base. On this machine, as on the previous one, everything was made of metal and no parts were built in plastic or polyester. His friend Ron Englert, who had already built three cars for Bob, took over the task, and not a single element was left in its original state. Everything was modified, reworked, replaced, or adapted.

This FXR won the first prize at the Oakland Show in 1995 and stimulated Bob Lowe's creativity; with his third motorcycle, called "Evil Twin," he pushed the limits of feasibility even further. He wanted to design a machine that abandoned the usual and allegedly compulsory central headlight, which appeared on all creations, even the craziest. By replacing the headlight with two stylized lights sunk in the front of the tank, not only did he achieve his goal but he also created an original front line that shaped the whole motorcycle by giving it the look of a shark muzzle.

All Bob Lowe's motorcycles are built on a theme and the design is thought through from the beginning. Those who manufacture

***253 top** This light belongs to Bob Lowe's first creation, presented at the celebration of the 90th anniversary of the Harley-Davidson Motor Company.*

***253 bottom** The first motorcycle Bob Lowe built on a Softail base was so successful, it encouraged him to continue his aesthetic flights of fancy.*

the technical parts are given strict instructions. The aesthetic part is up to Ron Englert, who uses a traditional method to shape all the metal parts: the hammer. He certainly has the necessary experience for this type of work, thanks to many years of practice in the car customization sector. As for the details, all parts are unique and are specially constructed for each machine. Moreover, since Bob knows a great deal about customizing cars, he can adapt these parts, or at least the ideas behind them, to achieve perfect motorcycles.

***254-255** After Bob Lowe's FXR won first place at the Oakland Show, his Softail (called "Evil Twin") won first place again – a unique performance for a motorcycle designer.*

***254 bottom** All the elements composing this machine are arranged to evoke the contours of a shark.*

***255 bottom left** The tank and the fork head form one single piece on this motorcycle.*

***255 bottom right** Bob Lowe's initial idea was to build a motorcycle that abandoned the traditional single central headlight. The result: a complete success in both looks and efficiency.*

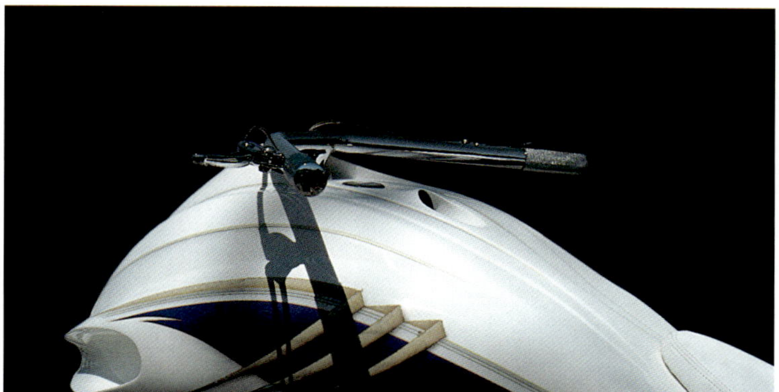

256 top A gentle line characterizes the rigid frame of this motorcycle. Its Evolution engine and a secondary transmission are covered by a finned Milwaukee Iron case reminiscent of old machines.

256 bottom This Evolution engine was equipped with a distinctive old-style crankcase cover and a gearbox.

257 top A Harley-Davidson fitted with a Shovelhead engine and refined to the max, for the look of an old racing motorcycle.

257 bottom This Knucklehead engine has been adapted to a motorcycle with a traditional chopper look, but special parts transform it into a modern machine.

Milwaukee Iron

The Milwaukee Iron company, already renowned in the United States for the quality of its work, has become increasingly popular in Europe for the same reason. Milwaukee Iron has deliberately chosen to do things differently from the others, thus creating its own style.

Milwaukee Iron manufactures parts and builds whole motorcycles. Their original projects generally keep the Harley-Davidson frame, transmission, and original engine. Sometimes, though, a frame, for instance, will be descreetly and carefully modified. In doing so, Milwaukee Iron tries to equal the quality of other professionals, but does so in a more modest way. Even when doing more radical transformations, though, they take the same approach. This, combined with their use of computer-driven lasers and machine tools, means that they can give their parts an irreproachable finish. This technique allows Milwaukee Iron to create unique parts, such as their famous air filter, which are then marketed through the catalogs of major distributors. The parts supplied by Milwaukee Iron are of exceptional quality, the best that American production can provide.

Their success is fully deserved: it has been achieved through an original method of working which looks to the future, and combines new techniques with outstanding craftsmanship to create truly world-class parts and unique motorcycles.

MILWAUKEE
IRON
Harley
Davidson
RevTech
RevTech

258-259 This Knucklehead with a rigid frame, shaped as a modern chopper, has a line that breaks radically from the style that ordinarily characterized it.

259 top This Knucklehead engine has been fitted out with Milwaukee Iron elements to improve its looks and with an air filter cover, which reproduces an old Harley-Davidson siren.

Genuine
MILWAUKEE IRON
Made in America
MILWAUKEE IRON

Paragon Locomotion

***260 top** Based on a Softail, this classic personalization combines the comfort of a good ride with the pleasure of an original motorcycle.*

***260 center** This Softail was customized with an eye to preserving original elements: engine, frame, fork, etc.*

***260 bottom** Thomas E. Worrell, Jr., head of Paragon Locomotion, a company specializing in the customization of Harley-Davidsons.*

***260-261** "Dragula", a Harley-Davidson with a body inspired by the vampire Dracula, built for a Florida customer.*

***261 top** The lines of this FLH have been improved for the exclusive use of Paragon.*

***261 bottom** This motorcycle may look like a restoration of an old Harley-Davidson, but it is in fact a modern machine, fitted with an Evolution engine and redesigned for a really retro look.*

The story of Paragon Locomotion, a company based in Florida, officially started in 1991, but its specialists have been working since 1984 within Angless Automobile Restoration, at the same address. At that time they only worked on cars, transforming some into rods and restoring others for collections. Nevertheless, they were keen on motorcycles, and using the skills and knowledge at their disposal, they personalized their own Harley-Davidsons. These were such a success, soon their friends asked for transformations as well. The phenomenon grew gradually, as the car market shrank slightly and the demand for customized motorcycles grew. This culminated in the creation of Paragon Locomotion, which specialized in the modification of Harley-Davidson engines.

The Paragon Locomotion team consisted of several people working in the workshop, an artist/designer, and a saddler specializing in leather. The saddler, a professional artisan, worked mainly on La Pera saddles, which he chose for their quality and their metal saddle base, and which he radically personalized every time. He also produced bike leathers, belts, clothes, and tank panels, as well as leather clothing for restored or personalized cars.

Regarding motorcycle manufacturing, the Paragon Locomotion method is always the same: the customer places his order, and two or three months later, he receives his motorcycle. All the personnel engage themselves on the project, with everyone giving his input, improving on other people's suggestions, and developing a detailed plan on paper of the final model, before any work starts in the shop.

The customers are people from all backgrounds, and whether they are doctors, police officers, or lawyers, they are selected customers: Paragon Locomotion manufactures only ten to twelve motorcycles per year. Given the number of employees, this provides the best possible rhythm for Paragon to carry out thorough work that is truly personalized and of top quality. It is true that they also execute additional small jobs on machines which have already been customized and whose owners merely wish to improve them in keeping with the new styles. On request, they carry out all imaginable styles of work, whether involving a modification of the frame or something simpler. On average, about 65% of the parts they use are unique, specifically created or modified for each motorcycle – a method which corresponds for the most part to the one adopted by the other customization professionals presented in this book.

262 top Paragon Locomotion modified this fuel tank to match the long line of this motorcycle.

262 bottom left All parts are engraved, even the tiniest: hence this brake caliper and disk, engraved to the same extent as the Springer fork.

262 bottom right The quest for quality finishing is pushed to the extreme. Even the least accessible part is decorated.

263 All mechanical elements and chrome-plated parts are engraved. Look at this engine in detail, for its sweep of arabesques.

HARLEY-DAVIDSON

265 top and 264-265 bottom ***The Springer theme, reminiscent of old Harley-Davidsons, appears here, but with Western-style paintwork.***

264-265 top ***A Softail modified according to the retro theme, with mud guards on a Springer fork and rear wheel.***

S&S
SUPER

266 This machine shows an Indian theme on a Softail Springer FXSTS base – a look entirely based on finishing and decor.

267 Chrome-plated and painted parts are combined in these exhaust outlets to create an Indian look. The cylinders are painted, the edges polished, and lower engine fins removed, and with space left free, the cylinder bases could be painted, too, according to the Indian theme. Although the air filter was an old S&S chrome-plated model which had not been modified, it was partly covered in embossed leather to match the motorcycle's paintwork.

268-269 This motorcycle has been completely disassembled so that all the parts can be chased or painted. During this process the steering caster angle was modified to lower the general trail.

Dave Perewitz

Cycle fabrication can be summed up by one name: Dave Perewitz. Dave Perewitz's relationship with motorcycles started in 1964, when he bought a Sportster. He was only sixteen years old, but he became interested in custom fabrication as soon as he started to work in a shop after school, sandblasting parts and doing simple paintwork. He got better, and then he dared to paint his Sportster by himself.

Friends recognized his talent and asked him to paint their motorcycles, too. What started as just an after-school job was turning into a profession. He studied mechanics, and later, he started working as a mechanic for a Chevrolet garage, still modifying motorcycles in his spare time. He took a crucial step in the early 1970s, when he opened his own shop to repair and personalize motorcycles. A few years later, he also opened a store, selling parts and accessories. But it was in 1975 that his career took the biggest turn. Dave Perewitz met Arlen Ness, who introduced him to the world of engine builders and, ultimately, helped Perewitz stand out among them.

By the early 1980s, Dave Perewitz had expanded his business and moved into his present workshop. Like every engine builder, he follows his own work methods. He starts with ideas, gathering and organizing them into a single design. Then his hands start moving, and finally he works directly on the concrete implementation of his project. Whether it is going to be an exclusive motorcycle or simply some flame paintwork –

one of his favorite touches – he works with equal spontaneity. As his business has grown, he has gathered together a select team who work to support his visions. Every year, Dave Perewitz unveils whole lines of stylized motorcycles, as well as machines built for certain customers, with paintwork that bears the distinctive Perewitz touch.

270-271 This type of FXR made Dave Perewitz known. He had been working for a long time, but he was known mainly on the East Coast until he started taking part in big meetings, presenting motorcycles with his kind of sophisticated paintwork. He then gained nationwide notoriety.

272 top **The Harley-Davidson Road King also interested Dave Perewitz. This first model, used by his son, was stripped and fitted with a new exhaust system and a few other modifications aimed to improve its performance.**

272 center **This more classic Road King customization preserves the elements of the machine while providing new paintwork and lightening the front, next to the fork.**

272-273 bottom **With this elegant figure, refined frame, and a look that evokes the Hamsters, the hand of Dave Perewitz on this Softail cannot be denied.**

273 bottom The Softail is one of Dave Perewitz's favorite frames. His paintwork always stands out from the rest of the machine. When he's finished, you can tell that he has designed a motorcycle for riding.

Wyatt Fuller, Ex-Razorback

Wyatt Fuller is a name long associated with customized motorcycles, created for years by Fuller's own Razorback Motor Works. Today, Razorback no longer exists, and Wyatt Fuller works for Harley-Davidson. Still, it is worth remembering the full history behind this king of customization from Florida.

Fuller was the founder, owner, and creative genius behind Razorback Motor Works in Pompano Beach, Florida. Early in 1991, his business began as a one-man custom shop in his home, but it quickly grew into the larger Razorback Motor Works.

***274 top** Known for machines with angular, massive forms, Razorback's custom work stands out. This full-bodied Softail epitomizes his creations.*

***274 center** Despite conventional lines, this Low Rider is a Razorback creation, customized to order with respect for classic design.*

***274 bottom** These machines, although less spectacular than others Wyatt Fuller has built, drew the attention of Harley-Davidson executives, making them decide to bring Fuller into the firm.*

275 top Characterized by less massive lines, this Softail shows elements that single it out as a Razorback motorcycle, especially the signed parts and the rear mud guard decorated with a spoiler.

274-275 Razorback transformed Softails by replacing original body elements with Wyatt Fuller parts. They weren't radical frame transformations, but they were exactly what Harley-Davidson looked for in products intended for the aftermarket.

In its short independent lifespan, Razorback racked up a good reputation and a strong hold in the custom field, first in Florida then in the rest of the U.S. and even in Europe, where some of its motorcycles still circulate today. In 1992, Razorback increased its custom and show market shares, thanks above all to customized Harley-Davidson models. Razorback style was distinctive: lines were more aggressive and gave the impression of controlled power. Razorback imposed a new style, which answered customer taste, and pushed Wyatt Fuller to the forefront in the field of customization. Never trained as a designer, still Fuller managed to envision new ideas and to make the most of state-of-the-art materials and techniques and excellent professional skills.

By 1994, Razorback's work was so highly acclaimed that Harley-Davidson bought the company and all of its assets. Harley-Davidson's Jerry Wilke, vice president of marketing for motorcycles and accessories, explained that the company snatched up Fuller, wanting his design work for its own.

While building custom masterpieces, Fuller didn't forget the prospect of building a business based on deluxe accessories. "We've now got about 22 custom components on the market under the Razorback name," he explained to *Big Twin,* the online Harley magazine, "and we're always working on more. We also do custom work on Fat Boy solid-disc wheels. People either send their wheels to us to have different slot patterns machined in them, or they can buy ones that we've already machined and plated. A lot of the bikes we build – including that white '92 Fat Boy – have wheels that have been modified."

Today, Wyatt Fuller contributes to the field of accessories and parts for Harley-Davidsons, through work that shows up in the official catalog. His Biker Blues, for instance, was originally a Harley-Davidson Fat Boy, entirely personalized and customized along the new jeans theme – an original customization that proves that the company from Milwaukee knows the commercial potential of such work on its motorcycles.

276-277 Abstract and futuristic paintwork take Harley-Davidsons far from earlier paintwork styles, mainly inspired by barbarians and scenes imagined by the artist Frank Frazetta.

X-Zotic

X-Zotic specialized in Evolution engine transformations, first with a kit that made the engine look like a Panhead, then by expanding on this look with bodywork parts. X-Zotic's biggest accomplishment, in 1995, was to reproduce a Knucklehead, using a Harley-Davidson Evolution. Engine and bodywork replicated that of an old Harley-Davidson, at least in aesthetic terms.

With its Evo-Indian kit, X-Zotic pushed the limits of feasibility even further, creating a machine with a pronounced Indian look on a Softail base. This model included all sorts of features that looked like those in Indian engines: cylinder head covers and a cam-cover with distributor, for example. The kit added exhaust pipes, metal mudguards, and footrests. It included either a front fork with headlight or the original fork bodywork, for those who preferred to economize. The carburetor pipe ran down the left side of the motorcycle with an Indian air filter and two Indian front and rear lights. In other words, the X-Zotic kit provided the elements that permitted a Softail to be turned into an Evo-Indian so well that it deceived a great number of amateurs.

***278 top** X-Zotic supplies elements that emphasize an old-style look, such as a crankcase cover or fork front bodywork, modern elements with a deceptively antique look.*

278 center and 278-279 Many Harley-Davidsons are fitted with these Panhead cylinder head covers, with no other retro transformation, simply to obtain a different engine look.

278 bottom X-Zotic produces parts kits to transform an Evolution engine into an old Panhead or Knucklehead, as is the case here.

279 bottom A Softail motorcycle, personalized with cylinder head covers made by X-Zotic to evoke the America of the 1950s.

Allen Ness

Battistinis

Battistinis, based in Great Britain, supplies Ness, Performance Machine, and Battistinis parts to several other European countries, adopting the American policy of building motorcycles as showcases. These motorcycles show a quest for aesthetic effects evoking the Arlen Ness style, but they also reveal Battistinis' mark. Company professionals work in total cooperation with Arlen Ness, and they always take into consideration that their motorcycles must cover several hundreds of miles. Of course, this is the best possible promotion for the parts and products their company supplies. Battistinis workshop is integrated with its distribution unit, which explains why all members of the company have a comprehensive knowledge of the products they sell and can continue to unveil the endless creative solutions to problems of motorcycle manufacture.

280-281 At first, Battistinis presented motorcycles either created by Ness or based on Ness lines, with parts coming from Ness' catalog.

282 top This Sportster, revised by Battistinis, typifies the harmonious forms designed to evoke a bigger Harley-Davidson than the original.

282-283 Battistinis rapidly adopted the policy of manufacturing new motorcycles each year to promote company products, in particular those by Arlen Ness. Like this FXR, they use Ness parts, but are built in England.

284 top ***OMP, the Italian specialist in aluminum parts, often combines its products with those by Carbon Dream, a specialist in carbon parts. OMP's originality has led to an international market presence for this company.***

284-285 ***TechnoPlus builds original parts with computer-assisted tooling machines, which enables state-of-the-art innovations in the Harley-Davidson custom field.***

OMP and Carbon Dream

OMP, an Italian parts manufacturer, makes items for Harley-Davidson such as risers, handlebar grips and controls, air filters, fork tees, rear-view mirrors, covers, pulleys, belt guards, and even wheels. The company has developed a following, especially in Europe, thanks to its high-quality parts, which have a special look which is different from American products. Motorcycles presented by OMP often include carbon elements, built by Carbon Dream, another Italian manufacturer that produces items for sport motorcycles, not only Harley-Davidsons but other brands as well. Cross-fertilization between the brands has meant that the ideas coming out of these companies are multiplied and adapted to Harley-Davidsons with very interesting results.

Zodiac

Based in Holland, Zodiac advertises that it builds parts for Harley-Davidsons, and it was actually the first European company offering parts for Harley-Davidsons on the European market. Its market plan was to supply both high-performance and custom parts for performance cycles and dragsters. In the mid-1990s, Zodiac started to showcase custom parts used to personalize Harley-Davidsons. The company thus showed to what extent these machines could be transformed using parts produced in Europe and Asia in addition to those imported from the United States.

285 top ***Zodiac, a Dutch company distributing parts built for Harley-Davidsons, also supplies custom and performance elements, thanks to a thorough understanding of dragster competition.***

285 bottom ***Tom's, a large German company, builds motorcycles to highlight the quality of the accessories and parts offered through its catalog.***

TechnoPlus

TechnoPlus is a French company specializing in precision parts, which dedicates its knowledge, techniques, and equipment to the service of the belle from Milwaukee. The courtship has proven fruitful. Company owner Claude Babot spurs on the development of new products, continually coming up with new parts instead of modifying those that already exist. Aluminum creations combine with original lines to provide parts which, once assembled, radically transform the Harley-Davidson aesthetic into something new. This philosophy, with its ever-new vision of possibilities for Harley-Davidsons, appeals not only to Europeans, but to Americans, as well.

Tom's

A special Softail was made for Tom's, a German company distributing products for Harley-Davidsons in Europe. It was built entirely with parts advertised in its catalog, and represents a promotional strategy similar to those adopted by American distribution companies. The basic idea for the machine came from the art department of Tom's. The machine itself was built by a customizing shop in Germany called Bike Schmiede, and it was designed to highlight the growing importance of the Asian parts market. The quality of products manufactured in Asia is constantly being improved, competing more and more strongly with other parts available from other countries. Built to match models conceived by great American designers, Asian manufacturing can reduce costs. In this case, the result is a clever combination of American state-of-the-art technology, German know-how, and Asian manufacturing expertise.

Photos courtesy Tom's

***286 top** This Softail is more classic in feel: a sophisticated look for a motorcycle intended for riding on the road.*

***286 center** The frame is fitted with a metal skirt similar to a spoiler; the rear mud guard consists of three metal parts molded to the frame, fitting out a "Hexco" oscillating arm with lightened lines. Engine cylinder heads and cylinders have been machined to the vertical, with three fins removed at the base of each cylinder. The result is an exceptional Canadian motorcycle.*

***287 top** A Canadian Softail, built with parts manufactured by the French company TechnoPlus: body, fork, oscillating arm, brakes, etc.*

***286-287** A traditional rigid chopper, but fitted with an Evolution engine, built entirely in Canada.*

Canada includes a good population of Harley-Davidson aficionados, and they are particularly keen on cycle transformations. Canadians are currently customizing the Harley-Davidson Low Rider and the Sportster, and several Canadian professionals produce such work. By contrast, the modifications carried out on FLH bases, whether they are Electra Glides, Electra Sports, or Road Kings, are still quite rare in Canada, except when a specialist like the painter simply known as "Bob," renowned in Quebec, becomes interested.

In spite of its proximity to the United States, Canada has developed its own particular line of transformed motorcycles. Whether they are massive machines or machines with slender, feline lines, they always suggest power – but they are never directly copied from the great American specialists. Canada's own sophisticated customization phenomenon has recently appeared at Bike Week in Daytona Beach.

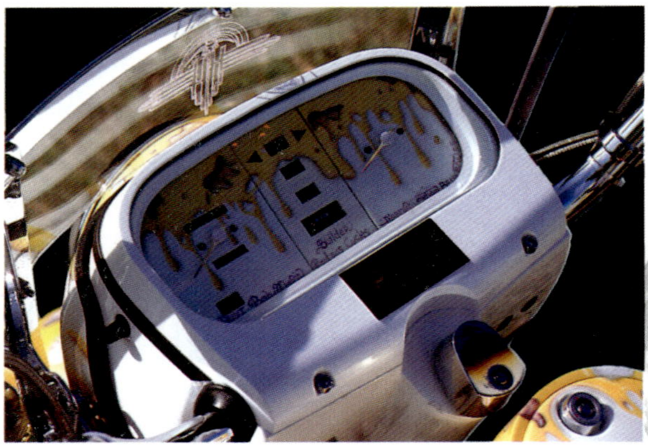

288 top left and center *A 1993 FLHS, entirely customized in Canada. Its owner, Michel Ethier, enlisted Deshaies Cycles' Bob and Sylvain Beliveau to build the machine. All the elements are painted according to the theme, even the inside of the dashboard.*

288 top center *The dampers are disguised under metal covers hand-built by Daniel Laugon.*

288 top right *The original motorcycle saddlebags are preserved, but their look is modified by extending the splashes of paint so as not to load the rear line.*

288 bottom *The original front fork is chrome-plated, while the motorcycle wheels are powder-coated.*

289 *Before being painted, this fuel tank was entirely molded so its line harmonized with that part of the frame on which its back was fixed. As a result, the tank seems to be one piece with the frame.*

Harley-Davidson
FLHS CUSTOM

290-291 This Harley-Davidson FLHS, modified by Michel Ethier and his American friends, won some prizes during shows at Daytona Beach in 1995.

V-Twin Engines and Harley-Davidson Models

V-Twin Engines and Customs

Motorcycles that are not Harley-Davidsons but look like Harley-Davidsons are increasingly frequent. First of all, several U.S. companies are building V-Twin motorcycles with aftermarket parts produced especially for Harley-Davidsons – parts such as frames, engines, gearboxes, or certain types of bodywork. Many Japanese motorcycles reflect the Harley-Davidson spirit, too, built as they are on the basis of V-Twin engines. As customer enthusiasm grows, and if customer demand remains unfulfilled, this sort of copycat growth will continue to happen.

Thus, the Harley-Davidson Motor Company has conceived a huge expansion plan, which should allow it to exceed an annual manufacturing quota of 200,000 machines by its 100th anniversary, in 2003. Harley-Davidson reached this goal and produced 291,147 machines for 2003, and 349,196 for 2006, the Harley-Davidson best year before the beginning of the difficult economy. However, such success may only continue to generate copies, as other manufacturers exploit the same touring custom motorcycle trend. There are certainly other great displacement machines, built with power and torque and an aesthetic that shows a strong personality, such was the Honda Walkyrie, equipped with a flat six-cylinder engine, and as the Yamaha Royal Star, provided with the old Venture engine of the same brand.

All these machines take their look from the custom-style, grand-touring conception, but they also provide an alternative to the Harley-Davidson, which has been a benchmark in this niche and has been monopolizing the market for years. These developments may threaten Harley-Davidson, since these motorcycles, unlike V-Twin imitations, may prove to have their own winning personalities. Luckily, the company from Milwaukee seems to be responding positively to the competition, thanks to the possibilities offered by customization and transformation.

The Prototypes

Today, other manufacturers copy Harley-Davidsons, but a few decades ago, exactly the opposite went on. At first, the Harley-Davidson look did not develop as much as did the spirit of the machines and a certain conception of their driving capabilities. Mechanics had to be reviewed and, at the beginning of the 1950s, the major preoccupation was the range diversification. Besides the big V-Twins, Harley-Davidson focused on the production of medium-displacement engines. Many sketches of small models similar to the British ones were drawn, and the company even thought of reintroducing the Harley-Davidson one-cylinder engines of the 1930s.

While slowly developing the Big Twins, Harley-Davidson contented itself with replacing the 750 WL with the K, then with the KH, and developing the 750 KL at the same time. The 750 KL, which had 150 cc less than the 900 KH, proved to be more efficient thanks to its carburetors, its chain distribution with overhead valves, and its aluminum engine, a concept completely new to Harley-Davidson. This motorcycle was not mass-produced, since marketing the K and the KH was already proving very expensive. Subsequently, in view of the huge success obtained by the first

Sportster in 1957, the KL project remained practically stillborn. Nine years later, the Harley-Davidson research and development department produced a scale model of a motorcycle called the X1000, with a four-cylinder engine and double overhead camshaft.

However, the X1000 required a huge investment, in terms of new buildings and machine tools, which made for slow progress. This was just as well, because the first Honda 750 soon appeared on the market. It was much nicer than the X1000, and if Harley-Davidson had tried to compete directly, the Honda could have struck a fatal blow.

At about the same time, the Harley-Davidson Motor Company was taken over by AMF. In 1971, the new company made an attempt to replace the Sportster with a 55-degree V-Twin 750 whose frame and tank were inspired by the XR, with exhaust pipes on both sides, an electric starter, overhead camshafts, and front and rear disk brakes. The new model was poised to compete against English and Japanese motorcycles, but Triumph and BSA stopped production and Japan's Kawasaki and Honda fought it out with each other, armed with the 900 Kawa and the 750 Four. Since that battle was already lost for Harley-Davidson, the 750 Overhead Cam was abandoned.

At the beginning of the 1980s, a new range of Harley-Davidson motorcycles appeared. Porsche engineers in Weissach (RFA) were charged with implementing a liquid-cooled V engine, whose displacement ranged from 500 cc to 1200 cc, divided between two, four, or six cylinders, according to the model.

Each engine had to be built with a double overhead camshaft, while still preserving the image of a traditional Harley-Davidson. The starting contract amounted to only $10 million, and as American engineers travelled back and forth to Germany, their discouragement grew. They ended up with a mediocre product, so disappointing that it was never even tuned up. They needed another $20 million to obtain a worthwhile result, so this project was also abandoned. Harley – Davidson will afterwards realize very persuasive production variations, in particular thanks to Buell, whose Harley-Davidson became the owner, and especially with the XXI century V-Rod.

***292 top** To evoke the old Panhead machines, this Heritage Springer is fitted with fringed saddle gear and a double exhaust system.*

***292 bottom** The Heritage Springer fork is fitted with a passing lamp, located above an imposing horn, which nicely covers the modern front damper.*

***293** A classic Evolution engine with a tank logo on top, which adds to the retro line of the machine, emphasizes the old style of the Heritage Springer line.*

Harley Davidson Evolution from 1984 to 1997

EVOLUTION BIG TWIN

In 1984, the 1340 cc Evolution engine appeared on some models. This machine represented a completely original conception, the Softail. The study of this new engine and its different designs started at the end of the 1970s, but the Sportster engine development program was rapidly upstaged by Harley-Davidson's preferred study on the new Big Twin engine, which it hoped to sell as soon as possible.

The Evolution engine included a newly designed combustion chamber, engineered in compliance with new government regulations aimed at reducing exhaust emissions. These new regulations came into force on January 1, 1984. The Shovelhead could not adapt to them, so the new Evolution engine proved to be vital. The same regulations pushed Harley-Davidson into manufacturing all its 1984 Shovelhead models in 1983.

The Evolution engine was a huge step forward for Harley-Davidson engineering. Not only did engineers create an engine in compliance with very strict regulations; they also achieved a level of reliability and performance unknown until then, also running on unleaded gasoline. Neither the valves nor the valve seats of the earlier Shovelhead could withstand unleaded gas. Given the new regulations, Harley-Davidson would not have survived without the Evolution engine.

In the Evolution engine, or the Blockhead, Harley-Davidson had an engine that could be easily adapted to all uses. It could fit on Softails with a chopperized look, like the Custom or the Springer; it could look as if it had come from the past, with the Heritage or the Nostalgia; it could be adapted to sport use on the road, closer to conventional motorcycles like Low Riders and Super Glides; and it would do the job required for deluxe touring with Electra Glides.

In 1991, Dyna Glide models emerged, built with a futuristic design generated by

294 top The Electra Glide was the first standard machine provided with electronic injection. It was tested in a limited version for a year before becoming an option on Harley-Davidson grand touring models.

computer . . . but, luckily, they were still based on the V2 Evolution. Future demands seem still to be answered by this engine, thanks to Harley-Davidson's effective study and research department. Since 1995, they have also brilliantly adapted Harley-Davidson engines to use electronic injection systems, in an Electra Glide limited series not yet available in Europe. New designs seem to be answering consumer taste as well, whether it's the 1995 Bad Boy or the 1997 Heritage Springer.

EVOLUTION SPORTSTER

In the middle of 1985, the 1986 Sportster with an aluminum engine appeared at a very low price – lower than the price of Sportsters with cast-iron engines. As Harley-Davidson had not produced any Sportsters in 1985, the concessionaires had no old stock to sell off and could therefore concentrate on selling the new bikes. The new Sportster allowed Harley-Davidson to recover financially.

With this model, Harley-Davidson offered a machine at a lower price than the Big Twins, which appealed to new customers. The plan was to generate future customers for the Big Twins, on the assumption that someone purchasing a Sportster now would be a future Big Twin customer.

Unlike the machines of other brands, Harley-Davidsons do not die; they are often reconditioned or transformed and continue to run or remain part of a collection, even when they become old. At first the Sportster was only available with 883 cc. Then, in 1986, a 1100 cc model came out to complete the range. These were not common displacement values for Harley-Davidson and the difference did not justify a higher price, according to the customers. Sales were moderate. Harley-Davidson reacted immediately. In 1988, the 1100 cc became a 1200 cc, which has been successful to this day. In 1992, all Sportsters were fitted with a secondary transmission by belt and, in 1996, two new models came out, the Custom 1200 and the Sport 1200, each targeting a very special group of customers. Both new Sportsters have bigger capacity tanks – 12.5 liter –, which have now been adopted across the 1997 Sportster line.

294 bottom When it appeared, the Softail Fat Boy was amazingly successful, prompting Harley-Davidson to try other variations on the Softail to differentiate them from existing models.

295 top The Evolution Sportsters developed rapidly, thanks to an attractive price and a good promotional campaign. Furthermore, new technology ensured reliability and higher performance.

295 bottom Sportsters are often transformed into racing machines and used by official riders of the brand.

The Harley-Davidson Evolution: From the End of 1998 to 2000

THE HARLEY-DAVIDSON TWIN CAM 88 FROM 1999 TO 2006

The 1998 Harley-Davidson line was characterized by the introduction of a new engine that would be mounted on future Harley-Davidson Big-Twins. In reality, 1998 was also the 95th anniversary of this Milwaukee company, but no new motorcycle was produced to celebrate the event, only a limited edition of a few models was available with special styling and paint: two-tone Midnight and Champagne Pearl paint with the 95th anniversary logo on the tank. These limited edition models covered the entire Harley-Davidson line: Sportster 1200 Custom, Dyna Wide Glide, Softail Fat Boy and Heritage Springer, Electra Glide Classic Injection and Ultra Classic as well as Road King Classic and Road Glide.

However, it is important to remember that in 1998 the Softail Night Train also made its first appearance. This model, based on the Softail Custom, but completely black, was destined for the European market. The tantalizing price made it such an immediate success that since 1999 it has been a part of the Harley-Davidson line. After the success of this model, Harley-Davidson recognized the demand for a basic Softail model and in 1999 introduced, or rather re-introduced, the Softail Standard. In reality, in 1998, aside from minor modifications to a few Big-Twins, only the Sportster XL 1200 S Sport underwent a substantial mechanical improvement. This enhancement was a revised ignition system in which each cylinder head was equipped with two spark plugs and new camshafts. Also the Sportster XL 53C Custom was introduced in 1998, with lots of chrome, a dual saddle and a rear disc wheel, and with the same specifications as the Softail Night Train, as well as a similar ride.

The real news, though, came in the summer of 1998 when the new Twin Cam 88 motor was announced, becoming the focal point of the Big-Twin Dyna and Electra models beginning in 1999, available at dealerships from the fall of 1998.

***296 top** New Twin Cam 88 motor. Available in two versions: aluminum tone or black paint.*

***296-297** Sportster 883 Custom XL 53C, also designed for the European market and later incorporated into the Harley-Davidson American range.*

***297 top** First version of the Softail Night Train with the 1340 cc Evolution motor. Originally to be sold only in Europe, it has now been incorporated into the Harley-Davidson American range.*

Therefore, officially starting with the 1999 models, Harley-Davidson equipped part of the Big-Twin line with a new engine called "Twin Cam 88" and nicknamed "Fathead." Although the exterior of this engine, with a 1450 cc capacity (to be more precise 1449 cc or 88 cu in), resembles a slightly more robust Evolution, the internal design and the technical aspects make it truly original. The relevant data speak clearly, with a torque of 106 Nm at 3500 rpm, higher than the Evolution, and a 14 percent increase in power (to 63 hp), making this a high-performance motor that was also more reliable and quieter. The V-Twin remained, in the Harley-Davidson tradition, at a 45° angle, and although it was only mounted on Dyna Glide and Electra Glide models in 1999, a similar design was used for the Softail beginning with the 2000 line. This engine, christened the "Twin Cam 88B" was unique in that it was equipped with twin chain-driven counter-balancers that reduced vibrations and could be mounted, without Silentbloc, directly onto the new Softail chassis.

Consequently, 1999 was the last year of production for the Evolution 1340 cc motor. Harley-Davidson did, however, launch the Custom Vehicle Operation, or CVO, program that produced two FXR Dyna Glide models specially equipped with Screamin' Eagle parts, that had unique styling and were built with an Evolution engine. This program continued in 2000 with a Road Glide mounted with a Twin Cam 95, which in reality is a Twin Cam 88 with the displacement increased to 1550 cc by simply increasing the cylinder bore. The program continued with the final production in 1999 of a motorcycle equipped with a 1340 cc Evolution engine, the 2000 FXR4 Screamin' Eagle.

***297 bottom** The following year, the Twin Cam 88B appeared on new Softail frames.*

A new model named the Dyna Super Glide Sport, designed with the Twin Cam 88 engine, was introduced in 1999. In 2000 the use of the Twin Cam 88B on Softails was accompanied with the introduction of the Softail Deuce model, a motorcycle with refined lines in which the tank, the seat and the fenders were modified and which also had a rear disc wheel.

In 2001 the Softail Fat Boy, Deuce and Heritage Classic were available in either a carburetor or fuel injection version for the European market, while in North America this alternative was available for the entire Softail family. Although the Dyna family was only

available with a carburetor, it was enhanced with a new model derived from the Super Glide Sport, the Super Glide T-Sport, that featured numerous adjustments to better configure the design to the needs of the driver. The model was equipped with nylon saddlebags and an adjustable steering column, minor modifications that once again announced that there were important innovations in the works at Harley-Davidson, as the centennial anniversary year approached. . . .

In the summer of 2001 the Harley-Davidson line for 2002 was presented, with the introduction of a new Sportster XL 883R racing bike, featuring styling inspired by the mythical XR 750 Dirt Track and which featured a 2-into-1 exhaust system. But the real innovation was found elsewhere, with the introduction of a new Harley-Davidson family: the V-Rod VRSCA. Harley-Davidson prepared well for its centennial celebration, an event that substantially increased demands for new motorcycles, which the company did everything possible to satisfy. This event, which coincided with the release of the V-Rod in 2002, explains the fact that other models saw only minor modifications without the appearance of truly new models in 2003, and that the entire line had logos and new paint dedicated to the anniversary.

2004 was the year of the Sportster. The Sportster family of motorcycles was simplified and reduced to four models, but benefitted from a redesigned engine mounted on Silentbloc, a new chassis and wider rear tires; in addition, the Sportster 883 and 1200 Custom were given a larger tank with a 17-liter capacity. The family then comprised the Sportster XL 883, the XL 883C Custom, the XL 1200R Sportster 1200 Roadster and the XL 1200C Sportster 1200 Custom.

The new FLHRSI Road King Custom was a harbinger of the future intermingling of the Harley-Davidson line. It is a motorcycle with a low rear suspension and wide handlebars that gives the classic Harley-Davidson a look designed more for excursions, combining the Touring and the Custom. That same year, the entire Electra Glide series was available in Europe only with fuel injection, while the FLHR Road King, FLHRS Road King Custom, FLHT Electra Glide Standard and the FLHTC Electra Glide Classic were still available with a carburetor option in the United States, and the FLHTCUI Ultra Classic Electra Glide and FLTRI Road Glide were exclusively available with fuel injection.

The Dyna T-Sport was discontinued from the line and all the other Dynas, FXDI Super Glide, FXDXI Super Glide Sport, FXDLI Low Rider and FXDWGI Wide Glide were available only with fuel injection in Europe and with a choice between fuel injection and carburetor in the United States.

The Softail Heritage Springer was also discontinued, while all of the Softail models, FXSTI Standard, FXSTBI Night Train, FXSTDI Deuce, FXSTSI Springer, FLSTFI Fat Boy, and FLSTCI Heritage Classic were available exclusively with fuel injection in Europe and with the option of a carburetor or fuel injection in the United States. The year 2004 was therefore notable for fuel injection technology, which was gradually added to all the Big-Twin families.

***298 top** Softail new model, the Deuce is moved by a Twin Cam 88B engine, that has been mounting on all Softail since 2000.*

***298 bottom** The first customized Harley-Davidson Touring series, the Road King Custom, is a low version of the Road King with a larger headlight nacelle.*

***298-299** The 2004 Sportster family featured a new frame and improved mechanics, but maintained the traditional 883 Standard with solo seat.*

***299 top** In 2003 the 100th Anniversary of Harley-Davidson was celebrated, and all models were available with special commemorative paint jobs, as appears on this Softail Deuce.*

***299 bottom** The 2002 and 2003 Sportster 883R models had a design that was directly inspired by the XR 750 racing bike, with a 2-into-1 exhaust system and a black-and-chrome engine*

Harley-Davidson
SEQUENTIAL PORT INJECTION
SOFTAIL®

The Softail family was enhanced with new models in 2005, adding the new Softail FLSTNI Deluxe that is reminiscent of Harley-Davidsons of the past and whose paint job was inspired by the 1939 models, and a new Softail FLSTCI Springer Classic that evokes the spirit of Harley-Davidsons from the late 1940s. At the beginning of 2005, Harley-Davidson produced a limited edition to celebrate the 15th anniversary of the Fat Boy, but this silver model with a black frame and the logo from the first Fat Boy was reserved for the American market.

The new Sportster XL 883L Low completed the family, offering a low riding position thanks to the lower front fork and rear suspension; however, it is more elaborate than the prior Hugger versions. The low has a single, narrow saddle placed slightly more forward than other Sportser models, with handlebars that extend further back and footrests that are also situated further back, all of which combines to make the position better for smaller drivers.

Typically the Harley-Davidson line for the coming year is presented the preceding July, with the new line entering the market starting in the fall. The Sportster XL 883R, however, was added to the line at the beginning of 2005, without being announced in July 2004.
The Sportster XL 883R was inspired by Dirt-Track models and had the orange color scheme of Harley-Davidson racing bikes, but with the engine block and transmission entirely in black. The Sportster line boasted at that time six models in the United States and five in Europe, as the XL 883L Low did not arrive in Europe until 2006.

Prior to 2006, the Harley-Davidson Big-Twin series had not undergone any noteworthy changes since the introduction of the Twin Cam 88 motor. However, in 2006 the first 6-speed transmission was introduced on the Dyna models. This family also benefitted from the addition of the FXDBI Street Bob that substituted the Super Glide Sport and from the FXDI Dyna 35th Anniversary, which commemorated the anniversary of the Super Glide. The other Harley-Davidson families also had new members: the Touring family added the FLHXI Street Glide, a GT type motorcycle with low lines and a similarity to the Custom (like the new Sportster XL 1200L Low that continued the concept of the XL 883L Low), while a new Softail model was introduced, the Heritage FLSTI.

The use of this new 6-speed transmission only on Dyna models seemed almost to be an experiment before adding it to the entire Big-Twin line. But the new air-pollution and noise-pollution laws in America and Europe will require further mechanical innovations by Harley-Davidson; the firm cannot simply curb the power and the noise of their motorcycles to the detriment of the performance and the pleasure of driving in order to meet ever more restrictive regulations.

300 Until 2006, the Softail Heritage maintained the mechanical design that was introduced in 2000 with the Twin Cam 88B and had the option of either a carburetor or fuel injection starting in 2001.

301 The 2006 FLHXI Street Glide was based on an Electra Glide Standard, with a lower line and more features, to create a Touring Custom that was designed to meet the expectations of many Harleyites.

302 bottom *The Sportster XL 883R, which was introduced in 2005, continues the racing styling of the XR 750, like its predecessors in 2002 and 2003, but does not have the 2-into-1 exhaust system.*

302-303 *The Dyna Super Glide FXDI 35th Anniversary model from 2006 commemorated the introduction of the Super Glide in 1971, and featured a special commemorative paint job for this motorcycle, but not an optional Boat-tail...*

302 top *A simple prototype that was presented in the fall of 2006, the XR 1200 turned out to be a success and became the first Sportster inspired by the XR 750 look to benefit from real mechanical improvements. Designed by American and European departments working in close collaboration, it has a decided sports character and will be a 2009 model that will be available starting in the summer of 2008.*

303 top *Beginning in 2006, the entire Dyna family, which included the new Street Glide FXDBI with ape hanger handlebars, featured a 6-speed transmission.*

Harley-Davidson Twin Cam 96 From 2007

The year 2008 marks the 105th year of Harley-Davidson's existence, and prior to commemorating this anniversary, the motorcycle manufacturer promised important developments. These changes were unveiled in the 2007 line, and were effectively unparalleled, as Harley-Davidson added their largest capacity engine ever to their line. The 1584 cc Twin Cam 96 replaced the old 1450 cc engine on all models of the Dyna, Softail and Touring families, but the other families were not forgotten. In fact, the V-Rod VRSC received two new models, and electronic fuel injection was added to all Sportsers, including the XL 50, which celebrated the 50th anniversary of its 1957 launch.

After the first aluminum Evolution 1340 cc engines (mounted on the 1984 Softail) were followed by the Twin Cam 88 1450 cc motor in 1999 and the Twin Cam 88B motor in 2000, Harley-Davidson reduced the timeframe for redesigning their engines by introducing in 2007 the Twin Cam 96 and 96B. These were very promising 1584 cc Big-Twins. For the manufacturer, it was the most important event since the launch of the VRSC V-Rod series in 2002. Unique in the recent history of Harley-Davidson, the introduction of this new technology was not gradual, but instead the engine was added to the entire Big-Twin line and thereby represented both a real gamble on the technology as well as a considerable investment in all fields, from research and development, to production, marketing, and finally communication.

The bet paid off, as the new Twin Cam 96 improved the acceleration of the motorcycles, while also making them easier to handle, even at the now higher speeds. Gear shifting is also improved with a smoother clutch, and the sound of the engine, not to be forgotten, maintains its full, harmonious character.

All of the 2007 Harley-Davidsons belonging the Dyna Glide, Softail and Touring series have the new engine, and also have the Cruise Drive 6-speed transmission that was used in 2006 on the Dyna Glide. In Europe, the air-intake system and exhaust system were specifically designed to meet the new air- and noise-pollution laws as well as increasing the power, the torque and the quality of the sound. The only Big-Twin series to receive a new model was the Softail, with the return of the FXSTC Softail Custom, while the FXSTS Springer and the FLST Heritage were discontinued.

In 2007 the Sportster series celebrated its

50th anniversary, and in anticipation of this event, Harley-Davidson introduced in July 2006 a limited edition of 2000 classically equipped Sportsters with a special 50th Anniversary logo, and the traditional small Peanut tank, to commemorate the debut of the model in 1957. At the beginning of 2007 the Sportster XL 1200N Nightster was added to the American line. The height of the saddle is identical to that of the XL 883L Low (25.3 inches or 643 mm). Equipped with a solo seat, slightly forward controls, wide handlebars, a black and chrome engine, as are all the non-mechanical elements, it has a license plate support on the left side, and directional signals that also act as brake, stop and parking lights.

However, the most important element of the 2007 line was the way tribute was paid to the heritage of the Sportster, while still looking toward the future, by equipping all Sportsters with the electronic fuel injection system ESPFI. To incorporate this new system, the Harley-Davidson engineers slightly modified the chassis and remodeled the back fender, the saddle, the tank, the exhaust system and the air intake system. The electronic sequential-port fuel-injection system is automatically regulated to provide a cleaner combustion, therby minimizing emissions as well as making the ride more enjoyable. At the same time, all the 2007 Sportster 883 models were redesigned to improve performance thanks to a new camshaft that increases torque at lower speeds.

The new Harley-Davidson Hands Free Security System, which is automatically activated and deactivated, was standard on all European models in 2007, and was available as an option in the United States. It is the only alarm system approved and guaranteed by the company. The Smart Security System is truly worthy of its name, thanks to its innovative characteristics and its ease of use.

In 2008, Harley-Davidson celebrated its 105th year of existence, and some models were enhanced with the optional 105th Anniversary limited additon package, which included a two-tone Copper and Vivid Black color scheme, an air filter box, a decorated leather saddle, and numerous other features. The optional package was available for the following eight models: FLHTCU Ultra Classic Electra Glide, FLHRC Road King Classic, FLSTC Heritage Softail Classic, FXSTC Softail Custom, FLSTF Fat Boy, FXDC Dyna Super Glide Custom, XL 1200C Sportster Custom, and XL 1200L Sportster Low.

304 In 2007 the 6-speed transmission was added to all the Big-Twin models, while the Softail family saw the return of a legendary model, the Softail Custom.

305 The Sportster XL 1200N Nightster from the 2007 American line goes back to the original spirit of the Sportster family, a pure, basic motorcycle that provides direct and immediate pleasure with an outstanding neo-retro look.

MOTOR
HARLEY-DAVIDSON
CYCLES
96 CUBIC INCHES
FATBOB
TWIN CAM

Harley-Davidson turned a new page in its history by discontinuing the mass production of the Springer forks that first appeared on Softail models in 1988, leaving them available only on CVO models. However, the radical new Softail FXCW Rocker was introduced in two versions: a base model and a Rocker C. The rear part of these motorcycles has typical Chopper styling, with a massive 240 mm-wide wheel situated in a Softail chassis which appears as though it has no suspension, like the old hand-built Choppers. Above the rear tire is a form-fitting fender, called a Rockertail, which mounts directly to the swing arm, without any visible support. While the Rocker has a solo seat, the Rocker C also has an optional passenger seat that attaches to the driver's seat by means of a removable support. A new Dyna, the FXDF Fat Bob, with drag bar handlebars having internal wiring, an exhaust system with a 2-1-2 collector with dual mufflers (as on the V-Rod), and a double headlight mounted on a large 49 mm front fork, was added to the line. These Softail and Dyna models demonstrate a pronounced Custom influence, in contrast to the Touring family which saw the discontinuation of the Road King Custom.

The Sportster family also reinforced Harley-Davidson's orientation toward the Custom, with the discontinuation of the XL 883R from the American line (although the European model remained), and the addition of the Sportster Nightster XL1200N to its offerings.

This was the American model from the previous year, adapted to the European market with a greater height from the ground thanks to the extended rear suspension. While the support for the license plate returns to the extreme rear of the fender, the directional indicators with the additional functions of brake, stop, and parking lights are retained.

The vitality of Harley-Davidson seems stronger than ever, with a line of Big-Twin and Sportster motorcycles that exhibit intelligent redesign and emphasize their differences with respect to the competition. Harley-Davidson offers a more classic product that often emphasizes presentation over other issues, even for the Custom and Cruiser motorcycles. By contrast, Harley-Davison entrusts the high performance motorcycles to Buell, the CVO models and above all the V-Rod series.

306 The new Twin Cam 96 and 96B engines that replaced the Twin Cam 88 and 88B on all the Big Twin models beginning in 2007 offer increased power and, more importantly, more character.

307 top The Dyna family, which is relatively classic with respect to the Softail, increased its customized models and, after the success of the Street Glide in 2007, added the Fat Bob FXDF with a wide front fork and two headlights to the line in 2008.

307 bottom Radical Chopper lines, with an elongated front fork and a rear fender that is attached directly to the swing arm, and that appears to almost touch the tire that it moves in unison with, distinguish the 2008 Softail Rocker, which has a decidedly Custom feel.

Harley-Davidson V-Rod from 2002

The year 2002 will go down in history as the most surprising for Harley-Davidson with respect to the innovations that gratified its public and, more precisely, the entire world of motorcycling. In effect, this motorcycle was primarily directed at a new clientele and, both to respond to the ever more restrictive air- and noise-pollution regulations, and to continue providing a high-performance engine, the V-Rod adopted a particularly modern design. This V-Rod can be summed up in a sentence: it is a Harley-Davidson with a Custom look and a Dragster feel. It was not by chance that the motor was christened "Revolution" as the design differs radically from previous engines produced by this Milwaukee company.

With the V-Rod, Harley-Davidson offers a motorcycle whose design and conceptualization the research and design groups had been working on since 1996. Stylistically, it is similar to a Low rider but with a pronounced aggressive character, with the engine highlighted in a way that transforms the motorcycle into a dragster – a name which it eagerly accepts and to which it fully conforms.

308-309 and 309 top The V-Rod was presented in the middle of summer 2001 and figured in the 2002 Harley-Davidson catalog. It is a unique engine with revolutionary features for a Harley-Davidson and is aimed at winning the Motor Company new customers.

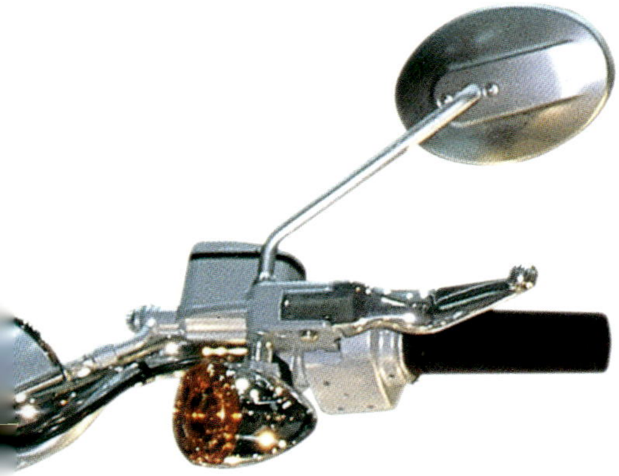

HARLEY-DAVIDSON

This first V-Rod model rapidly divided into different versions based on the Revolution engine, which in some models gained extra hp and in others was modified to a sport Dragster version. The work done by Harley-Davidson on the V-Rods benefitted both the aesthetic and the mechanical aspects by offering different features depending on its intended purpose: road use, urban use, or country use. With this new engine, Harley-Davidson targeted a niche clientele by offering vehicles customized from the very beginning, similar to the logic of the CVO program.

From the time of its introduction, the V-Rod family developed gradually as a complement to the Twin Cam 88 and Sportster models without having performance lowered by excessive limitations to noise- and environmental-pollution regulations, and thereby opened a new direction for Harley-Davidson. While it was impossible for the manufacturer to abandon the the type of motorcycle that guaranteed it the majority of its sales, it still needed to conform to these new restrictive laws. The solution rapidly appeared in the form of a new classic motor, the Twin Cam 96, that was used in the Big-Twin models starting in 2007, while the V-Rod family continuted with its own diversification, which began in 2004 with the VRSCB V-Rod that joined the VRSCA V-Rod. This VRSCB was merely a basic V-Rod with a black chassis, different trim and a more appealing price.

Subsequently, a VRSCR Street Rod was introduced at the beginning of 2005 in the United States, then soon after in Europe. This V-Rod with a straight exhaust system has 5 hp more than the VRSCA and the VRSCB, and is equipped with a 3-disc brake system and Brembo brake calipers. In 2006 the family still comprised three models, but the VRSCB was replaced with the VRSCD Night Rod, taking the straight exhaust pipes of the Street Rod and making a few other improvements that unleashed 120 hp. All of the V-Rods were equipped with a new chassis and a fuel tank that held 1.5 gallons (5 liters) beneath the seat, bringing the total fuel capacity up to 5 gallons (19 liters), which was done in response to one of the primary criticisms when the V-Rod was introduced: the lack of autonomy.

While 2007 was marked by the introduction of the new Twin Cam 96 engine, the V-Rod family was not neglected either; it boasted five models and benefitted from significant modifications. The base model, the VRSCA V-Rod, became the VRSCAW V-Rod, where the W refers to the massive 240-mm wide rear tire. The new VRSCDX is a Night Rod Special featuring a black color scheme and the same large rear tire of the VRSCAW, a modificiaiton that neither the Night Rod nor the Street Rod received. A more radical change was made to the new VRSCX, whose engine capacity was increased from the 1130 cc of the other V-Rods to 1250 cc by simply enlarging the cylinder bore and leaving the stroke unaltered. This VRSCX, which had the distinct Dragster "Screamin' Eagle/Vance & Hines" racing colors, was produced in a limited series of 1400 motorcycles.

While the 2008 V-Rod family once again comprised just three models, it is important to point out that they were all equipped with the new redesigned 1250 cc engine that improved both efficiency and driving comfort. Among the discontinued models was the VRSCX, a limited edition bike that was therefore eliminated from the line; the Street Rod, which was no longer available in either the American or the European market; and the basic Night Rod which remained available only in America.

Prior to being mounted on the limited edition VRSCX in 2007, this larger capacity V- Rod engine had already been used previously in two special models, the first in 2005 and the second in 2006, both of which were part of the CVO or "Custom Vehicle Operation" program.

310 The 2006 VRSCD Night Rod featured a new chassis that allowed for an increase in the size of the fuel tank and gained a bit more hp by adopting the Street Rod's exhaust system.

310-311 In 2007 the basic Night Rod VRSCA gave way to the VRSCAW that featured a much wider rear tire, which was designed to satisfy the expectations of V-Rod enthusiasts.

311 bottom The Night Rod Special VRSCDX was enhanced with numerous black elements, featured the new 240mm-wide rear tire and boasts an aggressive urban Dragster design.

Harley-Davidson

V-ROD

312 top The VRSCR Street Rod was, beginning in 2005, the sport version of the V-Rod with slightly increased hp, inverted front fork and triple-disc brakes.

312-313 The VRSCX 1250 cc, with the racing colors of the V-Rod Dragster "Screamin' Eagle/Vance & Hines," was produced in a limited series of 1400 motorcycles in 2007.

Harley-Davidson Evolution CVO Screamin' Eagle Program

Behind the name "Screamin' Eagle" lie genuine high performance motorcycle components, inspired by racing bikes and built in the pure spirit of adrenaline driven power. These components give Harley-Davidson motorcycles a performance on par with racing bikes without, of course, exceeding regulations for riding on public roads: a description that Harley-Davidson uses for these components that enhance motorcycles produced in its CVO program.

Although Screamin' Eagle motorcycles are known primarily as street bikes, it is important to realize that they were created in a reasearch laboratory specializing in high performance and that Harley-Davidson, which sponsors numerous Dragster race drivers and racing bikes that have Screamin' Eagle parts, has won many victories with Dragster "Screamin' Eagle/Vance & Hines" motorcycles. Screamin' Eagle is therefore a product of the research and production department of Harley-Davidson that specializes in high performance machines.

Strengthened by this experience, the

***314 bottom right** In 2003 and 2004, the FXSTDSE 1550 cc (95 cu in), a model based on the Deuce, was the first Softail CVO manufactured.*

***314-315** In 2000 the Road Glide CVO with the Twin Cam 1550 cc (95 cu in) engine was produced in a series of 1550 motorcycles, and in 2001 in a series of 1750.*

***314 bottom left** The 2003 Road King CVO experienced an increase in engine capacity to 1690 cc (103 cu in) from "just" the 1550 cc (95 cu in) of the 2002 model.*

Milwaukee manufacturer decided to produce motorcyles christened CVO Screamin' Eagle in a limited edition, starting in 1999 with the Custom Vehicle Operation program. The first model produced was basically a customized Dyna Super Glide FXD, and this FXR2 was produced in a limited series of 950 motorcycles featuring a single paint color and another 950 FXR3 with two colors, both of which retained the original engine capacity of 1340 cc.

In the following year, 2000, the CVO program continued. Starting again from a Super Glide FXD, the FXR4 was created with two-tone styling and a 1340 cc Evolution engine, that was produced in a limited edition of 1000 bikes. The Screamin' Eagle also introduced innovations by offering the FLTRSEI based on the Road Glide with a 1550 cc capacity Twin Cam 95 engine, offered in a limited edition of 1550 bikes.

The CVO concept, initiated cautiously, began to look promising due to the success of these motorcycles. While their price was somewhat higher than the motorcycles in the classic line, the inclusion of the Screamin' Eagle parts, the special paint jobs, and their polished appearance justified the price difference for these high-end, exclusive Harley-Davidsons. The Road Glide, the first model to benefit from real mechanical improvements other than increased engine capacity, laid the foundation of the CVO program which from that point on combined high performance with exceptional finishing details.

The Road Glide, a member of the Touring/Electra Glide family, entered the Harley-Davidson line in 1998 with a 1340 cc Evolution engine, and was later equipped with a Twin Cam 88 1450 cc engine and fuel injection in 1999. This motorcycle replaced the Tour Glide, discontinued that same year, from which was borrowed the idea of a large Harley-Davidson with bodywork attached to the chassis of the motorcycle. The only notable difference was the stylistic change with respect to the Tour Glide; the lines of the ever-present imposing bodywork were slimmed down by maintaining the polyester saddlbags but eliminating the Tour pack.

Custom Vehicle Operation used the Screamin' Eagle Pro Stock Racing machines, dragsters, as a source of inspiration for the production of the Road Glide FLTRSEI. This motorcycle was equipped with a high performance 1550 cc Twin Cam engine in chrome and silver, the increased capacity of which was obtained by enlarging the cylinder bore from 95.3 mm on a standard Road Glide to 98.425 mm on the Screamin' Eagle model, while leaving the stroke unaltered. The modifications that accompanied the incease of the engine capacity were accomplished with Screamin' Eagle parts that were expressly designed for legal street use. The kit included increased cylinder bore size, flathead pistons, special cams, a change in the calibration of the air intake and a different air filter. These modifications augmented the power by 10% and the torque by 14%. This Road King Screamin' Eagle included parts and accessories with a value in excess of $7500, as well as an exceptional paint job.

In 2001 the Road Glide CVO program was continued with the FLTRSEI2, of which 1750 were produced, and the FXDWG2, which replaced the FXR4 and was based on a Dyna Wide Glide but with a 1450 cc Twin Cam 88 engine, was produced in a limited edition of 1700 bikes.

In 2002 the Dyna Wide Glide FXDWG3 still had the same engine capacity, and was available in a limited edition of 1950 motorcycles, while the Road King, replacing the Road Glide, had a 1550 cc Twin Cam engine, and was produced in a limited edition of 2200 bikes, with the designation FLHRSEI.

In 2003, the year of commemorative models for Harley-Davidson Company, the Screamin' Eagle proposed for the first time a CVO model based on the Softail, the FXSTDSE Deuce, equipped with a 1550 cc Twin Cam 95 engine and available in a limited edition of 2950, and once again offered a Road King, equipped with a 1690cc Twin Cam 103 engine, and produced in a limited edition of 3200 motorcycles.

The CVO program has been extremely successful and reveals that the public is predominantly seeking high-performance motorcycles. In 2004, the FXSTDSE Screamin' Eagle was built starting from a Softail Deuce, with a 1550 cc Twin Cam 95 engine, the Road King was discontinued, and the new FLHTCSE Screamin' Eagle based on an Electra Glide was introduced, whose engine is a long stroke 103 cu in (1690 cc) model with a mixed air/oil cooling system. In the following year, 2005, Harley-Davidson produced three CVO models in response to the growing demand, with one model based on the V-Rod. The VRSCSE Screamin' Eagle, based on a V-Rod with the engine capacity increased to 1250 cc, was produced in a limited edition of 2400 motorcyles; the FLSTFSE Screamin' Eagle, based on a Fat Boy with a 103 cu in Stroker (long stroke) engine, was produced in a limited edition of 3400; and the FLHTCSE2 Screamin' Eagle, based on an Electra Glide Classic but with the engine capacity increased to 103 cu in (1690 cc), of which 3700 were produced. The CVO motorcycles were no longer being produced sporadically and in small quantities, and although they continued to be offered in limited editions, they became an integrated part of the Harley-Davidson line starting in 2005.

The CVO family, no longer confined to a limited availability, retained three models in 2006. The FLSTFE Screamin' Eagle continued to be based on a Fat Boy base with a 103 cu in engine, the FLHTCUSE Screamin' Eagle was based on an Electra Glide Ultra Classic with a 103 cu in engine, and the VRSCSE Screamin' Eagle continued to be based on a V-Rod with the engine capacity increased to 1250 cc.

***316 top** The 2005 CVO Fat Boy FLSTFSE was the first Softail equipped with a Stroker (long stroke) 1690 cc Twin Cam 103 engine.*

***316-317** The FLHTCSE Screamin' Eagle, based on the Electra Glide Classic with a long stroke 103 cu in (1690 cc) air and oil cooled engine, replaced the Road King as the base Touring for the CVO program in 2004 and 2005.*

***317** The success of the CVO line is growing and the models are being exported in larger numbers. They are available in different motifs and colors, as can be seen in this 2006 103 cu in FLSTFE Fat Boy, available in three highly stylized versions.*

Harley-Davidson

318 top In 2005, Harley-Davidson produced the first CVO based on a V-Rod, the VRSCSE Screamin' Eagle, with the engine capacity increased to 1250 cc.

318-319 bottom and 319 top ***The 1250 cc CVO VRSCSE, based on the V-Rod, which was available in various colors beginning in 2006 and had an increased engine capacity, was very popular and attracted many motorcyclists.***

HARLEY-DAVIDSON

320 top The high-end CVO FLHTCUSE, based on the Electra Glide Ultra Classic had a 6-speed transmission and the new Twin Cam 96 engine increased to 110 cu in (1800 cc). In 2007 it added to its 2006 features by including a GPS navigation system, but also experienced an increase in price to $33,495, taxes excluded.

320 bottom Both a high-end and high-performance vehicle, the new FLHTCUSE CVO 2006 Touring model was based on the Electra Glide Ultra Classic with an increased engine capacity of 103 cu in (1690 cc), leather luggage box and a Harman/Kardon AM/FM/WB/CD-MP3 with CB audio system from the Touring models, at a price of $31,995 excluding taxes.

The 6-speed transmission and the new Harley-Davidson engines were introduced in 2007, benefitting the CVO models and allowing for a consistent increase in the engine size, leading Harley-Davidson to offer four models: the FXDSE Screamin' Eagle based on a Dyna with a 110 cu in engine, the FXSTSSE Screamin' Eagle based on a Softail Springer with a 110 cu in engine, the FLHTCUSE2 Screamin' Eagle based on an Electra Glide Ultra Classic with a 110 cu in engine, and the FLHTRSE3 Screamin' Eagle based on a Road King with a 110 cu in engine. The engines for these motorcycles are the new Twin Cam 96 whose 1584 cc capacity was increased to 1800 cc (110 cu in), thanks to an increase in the cylinder bore.

For the 2008 CVO Screamin' Eagles, Harley-Davidson again produced four motorcycles, equipping the FL models with the Harley-Davidson ABS system. The FLHTCUSE3 based on the Electra Glide Ultra Classic, the FHRSE4 based on the Road King, the FXDSE2 based on the Dyna, and the FXSTSSE2 based on the Softail Springer, all maintained the Twin Cam 110 and 110B engines that they had in 2007.

The CVO models benefit from the increases in engine capacity to gain each time additional cubic centimeters and continue to set themselves apart from the base models. However, it also appears that these models serve more or less as guinea pigs for Harley-Davidson, which thereby has plenty of time to test the improvements that future generations of the standard Harley-Davidsons will receive.

320-321 and 321 bottom With the 2007 FXSTSSE, the Softail Springer benefited for the first time from the CVO program. Available in three colors, it featured a 110 cu in. (1800 cc) engine with counterbalancers and a chassis in the same color as the bodywork.

322-323 The 2007 CVO line was well balanced, with four motorcycles coming from the Softail, Dyna and Touring families, which in addition to the Electra Glide Ultra Classic, included the Road King that made its return with the FLHRSE equipped with the new 110 cu in (1800 cc) engine and a 6-speed transmission.

HARLEY-DAVIDSON

V-Twin Harley-Davidson Practice

The Harley-Davidson V-Twin engines have a long and complicated history. They can easily be identified by certain distinguishing characteristics. The names of the various Harley-Davidson V-Twins refer primarily to the shape of the cylinder head covers.

In order not to confuse them, it is necessary to consider the top part of the engine to establish its name and period.

FLATHEAD

The Flathead, with its finned and relatively flat cylinder-head cover, is available with a complete range of displacements: 750 cc, 900 cc, 1200 cc, and 1340 cc.

The most widespread model in Europe and particularly in France is the WLA 750, which arrived with the Americans during the Second World War. Built from 1929 to 1951, this 750 cc Flathead was a standard model with side valves, from which arose its common name, Lateral.

The official names for this machine were DL for the standard, DLD for the high-compression version, and simply D for the version with sidecar, until 1936. In 1936, the military versions of this famous 750 cc standard gained the reference tags WL and WLA.

A sport version called WLD and a racing version called WLDR were also derived from the same model.

As can be observed, the Flathead 750 was used in all imaginable ways. From 1932 to

1974, it was used to equip Servi-cars, three-wheeled vehicles built by Harley-Davidson primarily for garage owners to repair or tow cars. Later, the Servi-cars were produced for the American government. There were various versions available: model G was the standard; GL was a more sophisticated version; GA was designed for the Army; and GE was fitted with an electric starter. They were like normal solo-machine engines with a gearbox that added a reverse gear.

The 1200 cc engines were also produced in various versions: V for the standard, VL with high compression, and VS and VC for those fitted with gearboxes especially intended for sidecars. The first versions of these motorcycles were quite disappointing, but Harley-Davidson rapidly solved the problems by increasing resistance. Finally, the reliability of this model made it the most successful among customers.

The Flathead 1340 was produced from 1936 to 1944 in two versions: the standard UL and the ULH, to be combined with sidecars. This highly reliable machine appealed to many users, but the Knucklehead and the Flathead 1200 surpassed it. The Model K, an ancestor of the Sportster manufactured from 1952 to 1956, also had side valves, putting it too in the Flathead category. At first the available models were the 750 K and the KR, the sport version. Various racing motorcycles were built on the KR base. In 1954 the K series was completed with a KH 900 and, in 1955, with a more sporty KHK.

KNUCKLEHEAD

The Knucklehead is easily recognized by the aluminum covers, shaped like clenched fists, which decorate the cylinder head. The first model of this engine appeared with a 1000 cc displacement, in a standard version E, a sports version EL, and a version with a special side-car gearbox, ES, in 1936. In 1941, the Knucklehead 1200 was created with a standard version called F, a sports version FL, and a special sidecar gearbox version FS. The Knucklehead faced a few reliability problems at the beginning of its career, but they disappeared once the lubrication system was modified. Then its performance became very impressive for the time. Its career came to a rapid end in 1947, however, when it was abandoned and replaced by the Panhead, whose internal works were more straightforward.

PANHEAD

The Panhead, with cylinder head covers shaped like an upside-down saucepan, appeared in 1948. It offered the technical advantage of being much lighter than the Knucklehead, since many of its elements were made of aluminum. Its internal lubrication system was less complex, and thus more reliable. A standard 1000 cc model was available, referred to as E, a sports model EL, and a special sidecar model ES. The 1200 cc version offered the same three basic models: F, FL, and FS. The Harley-Davidson Police developed on a Panhead base, and when the Duo Glide turned into an Electra Glide in 1965 the Panhead gradually took on a GT machine aesthetic. Over the years, as the FLH, FLHF, FLE, FLF, and FLHB models emerged, the line became quite extensive, which explains the great number of Panheads still in circulation today.

324-325* *Throughout the life of the Panhead engine, transformations and technical innovations kept evolving. The Panhead first appeared in 1948 on Harley-Davidson bikes fitted with Springer forks. It equipped the Hydra-Glide models provided with telescopic forks from 1949 to 1957. From 1958, these Harley-Davidsons were called Duo-Glides, until, in 1964, the Big Twins adopted a rear suspension system. The Panhead's career ended in 1965, with a motorcycle called the Electra Glide, although its lower engine was incorporated into the first Shovelhead engines.

***326** Beginning in 1999, the Twin Cam 88 engine brought a breath of fresh air to the traditional 45° Harley-Davidson engine with air cooling. More powerful, it readily accepted fuel injection and offered both a smoother ride and better performance compared to the Evolution.*

***327 top** Thanks to Electra Glides fitted with Shovelhead engines, the Harley-Davidson Motor Company stood out for its deluxe grand touring motorcycles at a time when no other manufacturer had penetrated this market.*

***327 bottom** The old-style Sportster engine, fitted on Harley-Davidsons from the 1970s, was supposed to counter the invasion of Japanese engines. The Big Twin belonged to a different market niche, with well-defined customers.*

SHOVELHEAD

The Shovelhead, named for the flat, shovel-like shape of its cylinder-head covers, appeared in 1966. Early models – later unofficially called Early Shovelheads – did not have the same look as the modern Shovelhead, since they were fitted with an engine base similar to that of the Panhead. Available with 1200 cc, it modified its engine base at the end of 1969, adopting the typical look of today's Evolution engine and losing any resemblance to the Panhead. The Electra came out with 1340 cc only in 1978, followed by the whole Big Twin line from 1979 on. The Shovelhead was produced until 1983 and sold until 1984, when it was replaced by the new Evolution engine, made entirely of aluminum.

BLOCKHEAD

The Blockhead 1340 cc engine, named for its massive look and for the way its engine seemed cut out on top, was first sold in 1983 under the name Evolution. By then, the cast iron of the old engines had disappeared, making way for the advantage of an engine made completely of aluminum. Harley-Davidson really wanted to break from its past and mark its revival with a completely new, infallibly reliable engine. As a result, marketing emphasized the name Evolution, not the less-futuristic name Blockhead.

Today, the company from Milwaukee is still developing this engine. Engineers continue to face greater challenges along that path, responding to increasingly draconian requirements for pollution and noise control, put upon them by certain states and also by European Union countries. The future seems too difficult to predict. An injection system was introduced in 1996 on several models, so now the purchaser can choose between classic carburetion and electronic injection.

THE FATHEAD, OR TWIN CAM 88

The Twin Cam 88 was designed on the computer and then tested over 2.5 million miles before being made available to the public. This approach has made it the most reliable, long-lasting and powerful engine in the company's history; it is characterized by an increase in the piston displacement (with a modified combustion chamber), a new design of the valve ducts and exhaust, a sequential ignition system, and increased air intake due to a new front-entry air-box. The bore/stroke ratio has also been altered from the Evolution's 88.8 x 108 mm to the Twin Cam 88's 92.25 x 101.6 mm. Harley-Davidson has therefore moved away from the long stroke engine toward a squarer ratio for greater reliability and a higher engine speed. But the most surprising modification is the replacement of the single camshaft used till now with dual camshafts that work in harmony with short, silent transmission chains. Another change is that the tappets are no longer visible as they have been placed inside cases. These improvements lower the bike's running noise. In addition, the 50% larger cooling finning makes the engine more reliable, the engine casing is sturdier and the flywheel has a larger diameter, thereby benefiting rotation, and the integrated oil pump has a new circulation system for better lubrication of critical points. Fundamentally, this Twin Cam only uses 18 of the 450 components used by the Evolution, from which it is further distinguished by the more voluminous head covers (which is why it is called Fathead), the greater capacity of the air filter and the more extensive finning.

After the introduction of the Twin Cam 88 B, equipped with twin chain-driven counter-balancers that enabled it to be mounted directly onto the rigid frame of the Softail without causing significant vibrations, electronic fuel injection was progressively added to all of the Big-Twin models over the years, and in 2006 the Dyna family was equipped with a 6-speed transmission. However, the lifespan of the Twin Cam 88 engine, which was introduced in 1999 models, was short in comparison with other Harley-Davidson Big-Twin engines, being replaced in 2007.

Harley-Davidson
XL-1000

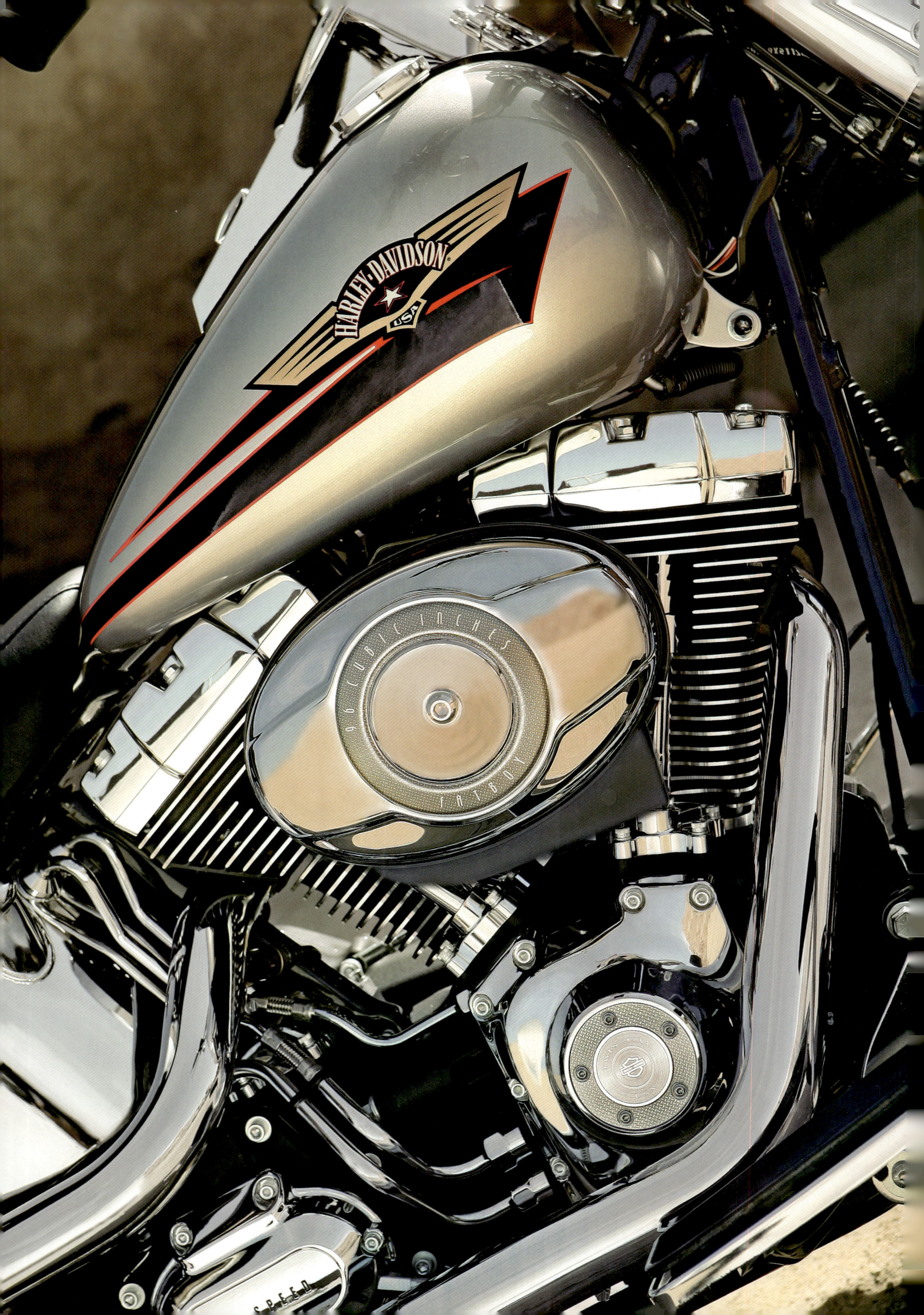
HARLEY-DAVIDSON
USA
96 CUBIC INCHES
FATBOY

***328** The year 2007 saw an important evolution for Harley-Davidson with all of the Big Twin models adopting a new engine and a 6-speed transmission. This Twin Cam 96 engine, which substituted the Twin Cam 88, was also available in the Twin Cam 96B version with internal counterbalancers that enabled it to be mounted directly to the rigid Softail chassis.*

***329** The electronic fuel injection system was a standard feature on all of the 2007 Big Twins. European models also featured an Active Exhaust System equipped with valves that enable the motorcycle to meet emissions standards by regulating the power, torque and exhaust noise based on the gear being used.*

THE TWIN CAM 96/96B

The new Twin Cam 96/96B, which replaced the Twin Cam 88/88B, was created to conform to the continually more restrictive anti-pollution regulations of various European countries and the United States. The new engine used decidedly more modern technology, surprising even for a Harley-Davidson, but applied it in an intelligent manner so as not to detract from the spirit of the motorcycle, and at times to even accentuate its character. Depending on the model, the new Twin Cam 96/96B 1584 cc engine supplies between 13% and 17% more torque than the Twin Cam 88 1450 cc engine that it replaces, which increases pick-up and acceleration and provides cleaner gear shifting. Harley-Davidson took advantage of this 9% increase in engine capacity and did further modifications to make this new engine smoother, more reliable and easier to operate than the Twin Cam 88. The engine speed is lower at similar velocities, as the speed of rotation of the 1584 is inferior to that of the 1450 due to the modified gear ratios of the new 6-speed transmission and the new Electronic Sequential Port Fuel Injection (ESPFI) system.

According to Skip Metz, manager of the Big-Twin Powertrain program at Harley-Davidson: "The Twin Cam 96 has been the biggest new-engine program in the history of Harley-Davidson."

Although this engine uses as its basis the motor introduced in 2006 on the Dyna models with 6-speed transmissions, there are more than 700 new pieces in the Twin Cam 96/96B. The only parts that were not modified were the cylinders, the rocker arms, rocker balances, and other items in the top part of the engine. The crankshaft, connecting rods, crankcases, transmission cases and transmission parts were all new. While the 95.3 mm cylinder bore of the Twin Cam 96 remains the same as that of the Twin Cam 88, the stroke length changed from 4.0 in (101.6 mm) to 4.38 in (111.1 mm) to increase the capacity from 1450 cc to 1584 cc, improving pick up and acceleration.
This engine has new electronic fuel-injection system, and the European models are equipped with an exhaust system that features valves in order to meet European laws, but which maintains as much as possible its original character. In fact, the Twin Cam 96 has up to 123 Nm of torque at 3400 rpm, while the Twin Cam 88 had only 109 Nm with the same speed, although the performance varies slightly depending on which model the engine is mounted.

The Twin Cam 96B, like the Twin Cam 88B, is designed to be mounted directly on the rigid support of the Softail models, and is therefore equipped with internal counter-balance shafts that absorb 90% of the principal engine vibrations.

The technological advancements of Harley-Davidsons equipped with the Twin Cam 96/96B engine push these motorcycles to the forefront. These advances include the ESPFI electronic fuel injection system, which receives information from a lambda sensor. This system was already used on many motorcycles produced by other manufacturers but makes its first appearance at Harley-Davidson in these quiet Big-Twins. The 2007 models have new fuel-injector nozzles placed at a 25° angle that offer better fuel atomization, an improvement in precision from the previous 8° nozzles, and a reduction of exhaust emissions. The active air intake system and the valve exhaust system, available only on European models, regulate the passage of exhaust gases according to the engine speed in order to optimize power, torque and sound while respecting European laws. This system, already used by many other motorcycle manufacturers, uses valves in the air intake and exhaust system that open and close in response to the Engine Control Module and the throttle position sensor. When the valves are closed, they reduce intake and exhaust noise, but improve torque, while when they are open the performance and the sound quality are improved due to the free intake of air and the improved passage of exhaust gases. Thus, Harley-Davidsons maintain their character while also satisfying air- and noise-pollution regulations.

The new oil pump offers a 10% improvement in flow and a 23% greater scavenging capacity, providing better lubrication and improved engine cooling, both are essential given the increased engine size. The 6-speed Cruise Drive transmission features helical-cut gear teeth which are quieter, and modified gear ratios which were changed, based on the torque curve of the Twin Cam 96/96B. The addition of a sixth gear permits an 11% rpm reduction in sixth, which at 75 mph (120 km/h) reflects a 368-rpm decrease, dropping from 3227 rpm with the 5-speed transmission to 2859 with the 6-speed transmission.

With this new engine Harley-Davidson has re-engineered the performance of the motorcyle by adopting decidedly new technology. Fortunately, this technology was applied discreetly, both visually and in terms of usage, and meshes with the whole to improve the performance of the motorcycle as well as the pleasure of driving without distorting its spirit, which remains the primary sales point for Harley-Davidson motorcycles.

SPORTSTER

The first true Sportsters, which followed the series K Flatheads, were manufactured from 1957 to 1985 and were christened XL. The power of the first 883 cc engines was similar to those of Triumphs, which invaded the United States and gradually eroded Harley-Davidson's market share. The 900 Sportster was manufactured with the base name XL until 1971 in different versions: standard, with electric ignition, and with various accessories. The XL 1000 cc replaced the 900 Sportster in 1972, and many versions of this model were made over the years, the two most important being the Café Racer and the XR. At the end of 1985 Harley-Davidson changed from a cast-iron engine to an aluminum one called Evolution, whose success continued to grow from that point on. Initially available in 883 cc, then in 1100 cc in 1986, the capacity was increased in 1998 to 1200 cc in order to better distinguish it from the 883 cc and also to increase sales. In 1992 all of the Sportsters replaced the secondary chain transmission with a toothed belt that all of the Harley-Davidson Big-Twins were already using.

Through the year 1995, the Sportster family did not enjoy much originality, and comprised the 883 and the 1200 models, which differed in the height of the saddle and the styling. However, starting in 1996, motorcycles directed at specific clients with completely different features appeared: the XL 1200C Custom and the XL 1200S Sport, followed by the XL 883C Custom in 1998. These models were an immediate success and Harley-Davidson continued in this direction in 2002 with the XL 883R, which has an XR 750 Racing design. Since 2001, all of the 883 motors increased power by 7% thanks to a modification of the exhaust system. At the same time, all of the XL engines received new pinions for the camshaft, a new oil pump, and also adopted the "Spinylok" cylinder liners of the Twin Cam 88, which in 1999 replaced the 80 Evolution engine. In 2003, the Sportser family did not change and continued to have four 883 cc models and three 1200 cc models. However, the family was dramatically altered in 2004.

The 2004 Sportser family was modified to include just four models: XL 883, XL 883 C Custom, XL 1200R Roadster and XL 1200C Custom instead of the seven models that were available in 2003. However, these Sportsters were completely new and marked an important step in the evolution of this Harley-Davidson family. The V-Twin Evolution engine, mounted on Sportsters since 1986, was redesigned, with improved engine cooling thanks to the resized fins and cylinder heads, and the new lighter pistons and oil nozzles that spray under the pistons. The connecting rods were also lightened and the maximum speed of the XL 1200 model engine increased from 5500 to 6000 rpm, thanks to the use of cylinder heads with large capacity and high performance cam shafts that improved torque. However, it was the motorcycle frame that underwent the most radical modifications, as the XL Evolution

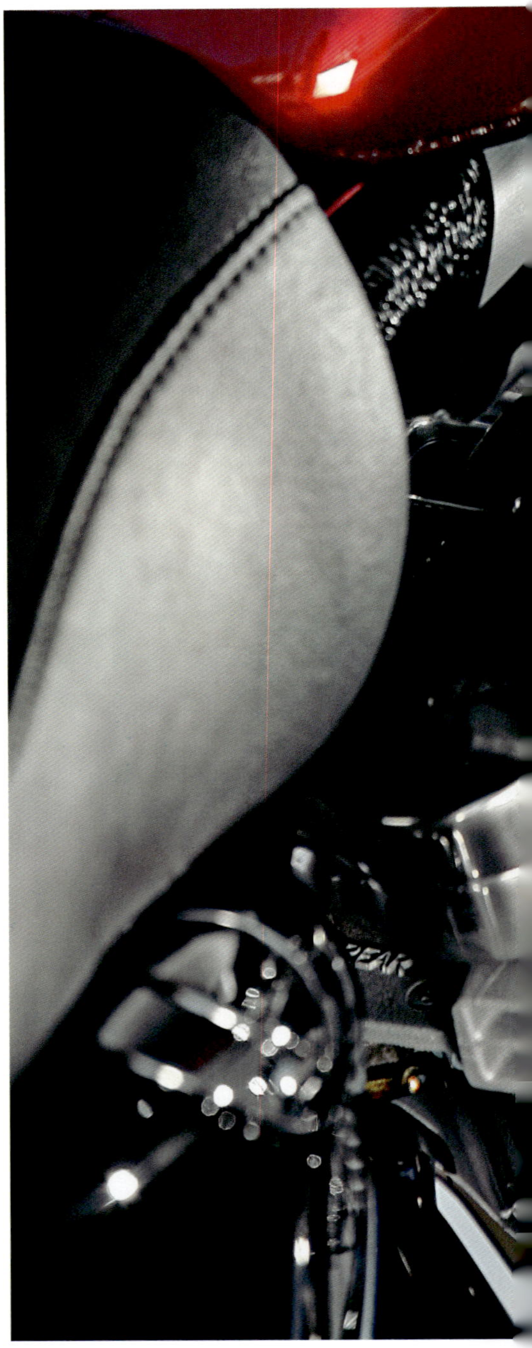

330 All of the Sportsters, which have an 883 cc or a 1200 cc engine capacity, and have a Sport, Custom or Standard orientation, feature electronic fuel injection beginning with the 2007 line.

engines were mounted on a new chassis by means of Silentbloc, which replaced the traditional rigid mounting of Sportster engines. This new mounting system decreased the vibrations felt by the driver and the new more rigid frame improved control and driving precision.

New models were integrated into this family over the years, like the XL 883R inspired by the XR 750 racing bike in 2005 and the XL 883 L Low (in 2005) that is based on the low Sportster Hugger, followed by the XL 1200L Low in 2006. In 2007 the Sportsters took another large step forward by replacing carburetors with electronic fuel injection. In addition, all of the Sportster 883 engines were redesigned to offer better

performance and to accept a new camshaft that generated greater torque at a lower speed. That same year, the family received a new addition in the USA with a Sportster featuring a radical look, the XL 1200 Nightster, which will be exported to Europe beginning in 2008. This model launched a new policy at Harley-Davidson for risk-taking design.

***330-331** Before being modified with a new chassis and an improved engine, the 2003 Sportster 100th Anniversary models were offered with a commemorative styling, as seen in this XL 883C Custom.*

***331 bottom** In addition to electronic fuel injection, the Sportster 2007 engines received a new camshaft and internal modifications that maintained high performance and improved the torque at lower speeds.*

THE V-ROD AND THE REVOLUTION ENGINE

The V-Rod is a Harley-Davidson having a Custom look and a dragster feel, with an engine christened "Revolution," as its concept differs radically from prior creations of this Milwaukee manufacturer. In addition, this new motorcycle is not produced at the Big-Twin assembly line in York, but in a new factory in Kansas City that was already producing Sportsters and Buells.

The Revolution is the first liquid-cooled engine available in the Harley-Davidson line, and features a V-Twin geometry with a 60° angle, dual overhead cams with four valves per cylinder and a permanent magnetic alternator. A true revolution, created through a collaboration between Porsche and Harley-Davidson, but which directly derives from the VR-1000 Superbike racing program. This new engine, which turns a one-piece forged crankshaft, has a capacity of 1130 cc and delivers a power of 115 hp at 8250 rpm, and a torque of 100 Nm at 7000 rpm. The dual overhead cams are driven by hydraulically tensioned chains requiring higher capacity heads with four valves, while the engine fueling is controlled by sequential port electronic fuel injection. Other innovations include the wet-sump oil pump, which denotes the absence of an external oil tank, and the hydraulically operated clutch.
The stylized large diameter 2-into-1-into-2 exhaust offers a high capacity for optimal performance, and decreases noise while maintaing the characteristic sound of a Harley-Davidson. An internal counter-balancer system and the mounting of the engine on Silentbloc reduces vibrations transmitted to the frame.

The silver perimeter frame is also new for Harley-Davidson. This concept offers better rigidity to the frame and, since aesthetics remain a key element for the brand, the curvature was hydroformed with no visible welding. The gas tank is located under the saddle, with the advantage of maintaining the mass close to the ground and optimizing the performance linked to fue injection. The dual-downdraft air intake system and the large air filter are found just above the center of the engine, in the position usually reserved for the gas tank, permitting an optimum direct flow of air to the engine. On the front, the headlight is outlined and surmounted by the new shell-shaped instrument panel with a tachometer and fuel level indicator with warning lights. The new instrumentation system also includes an auto-check with diagnosis upon ignition. The entire concept of the V-Rod clearly moves away from the tradition of the Big-Twins and brings it closer to the spirit of the Sportsters.

In 2006 all of the V-Rods were given a new frame and, in 2007, the VRSCAW V-Rod, VRSCDX Night Rod and VRSCX featured a 240-mm wide rear tire, while the new VRSCX boasted an engine capacity of 1250 cc by increasing the cylinder bore and leaving the stroke unaltered.

However, the real revolution arrived in 2008, when the new V-Twin Revolution engine was mounted on all of the VRSC models, with 123 hp and 11.5 daNm of torque. The engine capacity rosefrom 1130 cc to 1250 cc and the VRSC features a softer clutch with a slipper-clutch system. The ball-bearing system reduces the torque of the clutch during engine braking, enabling down-shifting without risking rear wheel lock-up and thereby avoiding skidding or loss of traction.

This family of Harley-Davidsons has found its own market which now allows for a diversification of the models by adding features designed to improve the performance for the specific intended use each model.

Harley-Davidson engines are designated bythe numbers 45, 74, 80, 88, and 96 (the most recent), which refer to the engine capacity, not the year of construction.

45 cu in = 750 cc
54 cu in = 883 cc
55 cu in = 900 cc
61 cu in = 1000 cc
74 cu in = 1200 cc
80 cu in = 1340 cc
88 cu in = 1450 cc
96 cu in = 1584 cc

These conversions are clearly not 100% accurate, but are the numbers currently used by Harley-Davidson to convert the different engine capacities of its motorcycles.

332 top The V-Rod's engine is not the only surprising aspect of this motorcycle, as the chassis also features unusual lines, even by the standards of competing manufacturers.

332 bottom The VRSCAW V-Rod replaced the base V-Rod in 2007, and featured a rear wheel that supports a 240 mm-wide tire, which was the primary modification that V-Rod owners had performed in previous years.

333 top Beginning in 2008, all of the V-Rod models, like this VRSCAW, adopt the new 1250 cc engine which replaces the 1130 cc engine that had equipped this line, with the exception of the VRSCX, since its first appearance.

333 bottom
The V-Rod engine was christened Revolution, a name that suits it given its characteristics: liquid cooling, 4 valves per cylinder, 60° V-Twin... partially derived from the VR 1000 race bike.

Harley-Davidson: A new era

A new century means a new era for Harley-Davidson, at least that seems to be the motto for the Milwaukee manufacturer whose centennial anniversary in 2003 appears to have gone to its head. The end of the previous century saw the introduction of the Twin Cam 88 engine, while in 2001 the company christened the revolutionary V-Rod, a completely new concept for Harley-Davidson,

but which also maintained the spirit of the brand. Also new was the introduction by Erik Buell of his surprising XB-9R in 2002, which was based mechanically on a Sportster, and which created a motorcycle with an original concept and a revolutionary frame that would later be used for all new Buell motorcycles. With 100 years of history behind Harley-Davidson, the year 2003 represented worldwide recognition of its status as a living legend in the field of motorcycle production. The truth of this can be seen in the changes that followed: the innovative Sporster frame in 2004, and the exceptional Twin Cam 96 engine in 2007, to mention only the most important developments.

Even a novice motorcycle enthusiast recognizes that the V-Rod is a Harley-Davidson different from the other models. It is still a Harley-Davidson and it shows; the tank even confirms it. However, the rest of the motorcycle does not resemble the usual products of this Milwaukee manufacturer. This model has completely different lines that clearly distinguish it from the Custom and Cruiser motorcycles available on the market. Long and very low, with its modern V-Twin engine, it looks like a personalized Custom model, and it is here that the strength of Harley-Davidson resides. With the V-Rod, the Milwaukee manufacturer has created a new standard that has rapidly become popular, thanks to the diversification of the models.

This same process was seen several years earlier when the Softail line, with its aluminum Evolution engine, progressively became the reference point for Custom motorcycles.

Initially, criticized and insulted by the competition and also by the majority of Harley-Davidson enthusiasts, this motorcycle with its modern engine immediately attracted a new generation of Harleyites and rapidly became a standard for the Harley-Davidson brand; it was later expanded to include various models. Today, the Softail line is copied by many manufacturers who, inspired by its characteristic design and V-Twin engine, want to produce a Custom-type motorcycle that includes the Softail frame technique, which is found on the majority of the Custom or Cruiser motorcycles of other brands.

The success of Harley-Davidson at the beginning of the millennium proved to be real. The objective that the company set in the mid-1900s was to reach a production of 200,000 motorcycles in 2003. This number was comfortably surpassed before this date, with the total production of Harley-Davidson in 2000 already at 204,592 motorcycles, increasing to 234,461 units in 2001, then 263,653 in 2002, and coming close to reaching 300,000 in 2003 with 291,147 Harley-Davidsons produced. It is worth noting that this number was almost 100,000 more motorcycles than originally projected. These figures should also be compared with the motorcycle company's almost non-existent production in the mid-1980s; specificially in 1985, when fewer than 35,000 motorcycles were sold worlwide.

At that point, in its 100th year of existence, the Harley-Davidson Big-Twin family was composed of the Softail, Dyna Glide and Touring, along with the Sportster family, which assured a regular and steadily growing sales volume and a faithful clientele. The introduction of the V-Rod repeated the success of the Softail program and gained the attention of both the motorcycle world and the general public with its unveiling of this new generation of motorcycles. With its liquid-cooled engine equipped with a modern four valve per cylinder technology, the V-Rod was capable of opening a new market composed of people who were previously hesitant to enter a Harley-Davidson dealership and to drive a bike considered to be different by both motorcyclists and non-motorcyclists alike.

334 left The Softail Standard was discontinued from the Harley-Davidson line at the beginning of the 1990s, but made a successful comeback in 1999 when it was re-introduced to the Softail family.

334 right The Softail Heritage Classic, based on a Softail and equipped with a Twin Cam 88B engine from 2000 to 2006, and thereafter a 96B engine, experienced continued success since its introduction to the Harley-Davidson line at the end of the 1980s.

334-335 The Dyna Glide Low Rider, one of the solid values offered by Harley-Davidson, was used as the basis for many subsequent models designed for brief street use and with a highly customized style.

335 top left The Buell Lightning X1, built with a Sportster engine, remains a traditional high performance model manufactured for sport bike enthusiasts and is designed to indirectly bring new loyal clients to Harley-Davidson.

335 top right The Buell XB 9R represents an exercise in style and, more importantly technical ability as far as the cyclist is concerned. It is the Buell of the future, whose revolutionary chassis is designed to accommodate a higher performing engine than the Sportster engine...

The reorganization of the production unit

To attain these 2003 production results and then to quickly surpass these numbers, it was necessary for Harley-Davidson to implement its production policy by reorganizing existing manufacturing sites and developing new ones.

- The Corporate Headquarters on Juneau Avenue in Milwaukee, Wisconsin, was the original seat of the company. The first factory was constructed there in 1906 and expanded over the years before becoming the famous red brick building which has now been converted into offices and houses the general and administrative branches of Harley-Davidson Company.

- Powertrain Operations on Capitol Drive in Wauwatosa, on the outskirts of Milwaukee, which was the factory where the parts for all Harley-Davidson models were produced, is now almost exclusively dedicated to Sportsters. Only a small unit for repairing Evolution and Twin Cam 88 engines is left from the Big-Twins as the factory now produces Sportster engine parts and assembles and checks Sportster, Buell and Blast engines. These engines are then sent to the motorcycle assembly plant in Kansas City for the Sportster and to East Troy for the Buell and the Blast.

- The Product Development Center in Wauwatosa, built in 1997 behind the factory on Capitol Drive, is the Research and Development Center for Harley-Davidson and also houses the Purchasing Department.

- Powertrain Operations on Pilgrim Road in Menomonee Falls, Wisconsin, to the north of Milwaukee, is a owned factory that has produced Harley-Davidson Big-Twin engines since 1996.

- Tomahawk Operations in Tomahawk, Wisconsin, continues to produce the plastic elements for Harley-Davidsons, such as bodywork, trunks, saddlebags, sidecars, windshields, etc.

- The P&A Distribution Center in Franklin, Wisconsin, holds the warehousing, packaging and shipping services for Harley-Davidson parts and accessories destined for dealers and distributers worldwide.

- Vehicle Operations in York, Pennsylvania no longer houses the Harley-Davidson Rodney C. Gott Museum. The space once occupied by the museum has been reutilized as an assembly plant and enlarged to meet increased production demands. However, it now only handles the production of a few parts, as well as the assembly and painting of the Softail and Touring models and their respective quality control.

- Vehicle and Powertrain Operations in Kansas City, Missouri is an ultra-modern 358,000 sq. ft (33,258 sq. m) factory, opened in 1998, that produces the mecchanical parts for the V-Rod, assembles and paints the Sportsters, Dynas (since 2002) and V-Rods, performs quality control, and corrects product defects.

- Harley-Davidson Assembly Operations in Manaus, Brazil is a unit that has assembled imported parts to produce Harley-Davidson motorcycles since 1998.

***336 left** It is in the Product Development Center, or Willie G. Davidson Center, that Harley-Davidson does the research and development for its future models and motorcycle parts.*

***336 right** In Kansas City, Vehicle and Powertrain Operations is the factory for the V-Rods, but also the location where Sportsters and Dynas are assembled.*

Harley-Davidson 100 years of fame

The centennial anniversary of Harley-Davidson was a worldwide event, with celebrations on every continent. In fact, while Milwaukee was the focal point of the celebrations, where hundreds of thousands of Harley-Davidsons converged from all the major cities on the American continent, the celebrations were not limited to the United States. Prior to the Milwaukee gathering in August 2003, numerous celebrations were organized in the USA starting from 2002 onward, called the Open Road Tour, and on every continent there was a celebration, starting from the spring of 2003.

The centennial celebration of Harley-Davidson paid homage to both the Harley-Davidson motorcycle and to the company itself, which has been able to manufacture modern vehicles while preserving its long tradition. The Milwaukee manufacturer has carefully cared for its own history while also enlivening its business by holding commemorative events in past years. Large gatherings were organized for important Harley-Davidson anniversaries: the 85th, the 90th, and the 95th in Milwaukee, which attracted an ever greater number of participants.

The Open Road Tour was organized for the company's 100th anniversary and was celebrated in ten cities throughout the world, spread over several continents. The tour lasted for a year, with the grand finale being held in Milwaukee at the end of August 2003. The opening ceremony was held 19-21 July 2002 in Atlanta, Georgia, followed by the next event on 16-18 August in Baltimore, Maryland, then 6-8 September in Los Angeles, California, then 27-29 September in Toronto, Canada, and finally 25-27 October in Dallas, Texas. The Open Road Tour then continued on to Sydney, Australia on 14-16 March, 2003, then Tokyo on 25-27 April, before reaching Europe, with Barcelona on 27-29 June representing Southern Europe and Hamburg on 25-29 July repesenting Northern Europe. The finale was held 28-31 August in Milwaukee. Many Harleyites participated in the Ride Home, which started in mid-August and originated from four different American gathering spots: the Northwest, the Southwest, the South Central, and the Northeast. The Ride Home finished in Milwaukee, where Harley-Davidson triumphed at the end of August 2003 with a celebration that attracted, according the the Milwaukee police, between 200,000 and

250,000 motorcycles and 500,000 people.

On 30 August the large parade, limited for security and time reasons to "only" 10,000 Harley-Davidsons, crossed downtown Milwaukee following a route indicated with Harley-Davidson centennial banners, to finish in West Maier Festival Park. A mixture of Harleyites, Milwaukee residents, tourists, and even entire families made up the crowd of people who came to see the parade and to admire the Harley-Davidsons. These motorcycles came from all over the world: the Harley-Davidsons of police and fire departments, those of the Harley Owners Group chapters from various states and countries, with their flags rippling in the wind, motorcycles of the management and employees of Harley-Davidson Company, and finally those of political figures.

In contrast to prior Harley-Davidson celebrations in which the parade was the highlight, this centennial parade did not mark the end of the ceremony. Instead it ended the following day on Sunday, 31 August 2002, at the official closing ceremony held near Veterans Park, on the shores of Lake Michigan. An enormous setting accommodated the event, which included speeches, performances, and concerts as well as the introduction of the 2004 Harley-Davidson line. For safety reasons, only the first 150,000 who had registered at the beginning of the celebration four days earlier were admitted. The tens of thousands of spectators who were excluded had to content themselves with watching the events from outside on large screens.

At the end of the celebrations, some Milwaukee residents could be seen holding signs with the invitation: "Come back in 2008, for the 105th Anniversary of Harley-Davidson."

Motorcycle families in constant evolution

Harley-Davidson had just celebrated its centennial with a record production of motorcycles. The demand, however, did not diminish and it was not long before production surpassed the 300,000 mark. In 2004, with a total production of 317,289, a 9% increase from the anniversary year of 2003, Harley-Davidson confirmed that the well-being of the company was not a temporary phase. The line was even enhanced with new Sportser, Big-Twin and V-Rod models that sustained this growth and contributed to Harley-Davidson's production of 329,017 motorcycles in 2005 and 349,196 in 2006. Also in 2006, the first Harley-Davidson dealership was inaugurated in China, which opened shop in Peking under the name Beijing Feng Huo Lun, and the logo FHL. In 2008 Harley-Davidson opened a new museum (previously in York) located in Milwaukee that contains a greater number of motorcycles for the 105th Anniversary.

All the Sportsters were equipped with fuel injection in 2007, while the Big-Twins were given a new generation engine, the Twin Cam 96, that replaced the Twin Cam 88. In 2008, was the V-Rod's turn to undergo a small revolution, since all the models were equipped with the new V-Twin Revolution engine, the 1250 cc engine of the VRSCX V-Rod that was produced in limited edition in 2007 with Dragster Screamin' Eagle/Vance & Hines V-Rod styling. To commemorate the anniversary of the company, some models were equipped, as has become the custom, with a 105th Anniversary package in limited edition.

Thanks to a line composed of families with very different attributes, any motorcyclist can find a Harley-Davidson that matches his or her character and desires. The question of age is not really an issue, as the sporty Buells are more suited to a younger clientele, as is the Sportster which is more spirited than the Big-Twins. The V-Rod attracted a new clientele to Harley-Davidson, but it also lured some motorcyclists who were already driving Sportsters. A slightly more detailed study of the different families, test-driving a V-Rod, an Electra Glide, a Dyna Glide, a Softail and a Sportser, and discovering the new universe of Buell products and the sportbikes from Harley-Davidson Company, allows for an appreciation of the diversity of options offered by the company that enables each person to find the most suitable motorcycle.

337 In Milwaukee at the end of August 2003, a large parade that was limited to just 10,000 motorcycles (for reasons of security) and lasted for hours, celebrated the 100th anniversary of Harley-Davidson.

Buell, roadster and Harley-Davidson Sportbikes

Since Erik Buell founded the Buell Motorcycle Company in 1983, he has worked exclusively with Harley-Davidsons as a base, in contrast to others who have diversified by using different motorcycles to make their own sportbikes.

Erik Buell, who was employed as an engineer by Harley-Davidson, developed a definite taste for sportbikes after participating in numerous speed tests in the 1970s. In 1983, he left Harley-Davidson and founded his company to produce a competition race bike. In 1987 he settled in Mukwonago, Wisconsin, located approximately twenty minutes from Milwaukee, to assemble his own approved race bike, the RR1000, which was completely covered in body-work and equipped with an XR 1000 engine. Since the RR1000 was lighter and better performing than the bike that the Harley-Davidson and the Harley Owners Group were racing, these particpants decided to use Buells.

The Buell posesses a chromium molybdenum tubular trellis frame from which the engine is suspended by means of "Uniplanar" system (Buell patent). The Uniplaner system is composed of Silentbloc and adjustable axles, which allows for centering the engine with respect to the swing arm to which it is attached. This system insulates the driver from the vibrations of the V-Twin. After the RR1000, Buell produced the RR1200 from 1988 to 1990, which was powered by a modern aluminum Sportster Evolution 1200 cc engine, purchased directly from Harley-Davidson. Unfortunately, although Buell designed and created high-quality motorcycle parts, he left the engine practically untouched and therefore could not seriously compete against the race bikes of other manufacturers. The RS1200, produced between 1989 and 1990, evidenced some technical improvements to the motorcycle, but more importantly comfort and practicality were improved thanks to the addition of a passenger seat. The saddle, which has a zipper, can change from a single to a double seat thanks to the turnable rear part that, once raised, supports the passenger's back. The Buell saga continued for a few years before the company needed a recapitalization in order to continue to doing business. Beginning in 1993, thanks to a capital contribution from Harley-Davidson to the Buell company, which had almost ceased activities, the production of racing bikes based on the Sportster could begin again and even diversify.

At that point Harley-Davidson held 49% of the shares, after having delivered an infusion of capital of approximately $1 million to the company. In 1994 production of the Thunderbolt S2 began, which was joined the following year by the touring-sport version, the S2T. The 1996 Lightning S1, the first roadster from Buell, was powered by an improved engine and experienced a great success until 1998, while the Thunderbolt was modified and became the Thunderbolt S3 and then the Thunderbolt S3T in 1997. Also in 1997 the Cyclone M2, a new model with a more relaxed driving position and a wider saddle was produced with the Sportster 1200 engine, like the other Buells.

338-339 This Buell RR1200 evolved into the RS1200, a model in which the driver's back rest can be raised to free the passenger seat, which then features a back rest.

339 Erik Buell manufactures motorcycles with a racing orientation, and it is enough to slightly modify them at a Harley-Davidson dealership to have high performance competition bikes built with Sportster engines.

338 The Buell RS 1200 was the first Buell to shed its ponderous bodywork in favor of more pleasing lines, although the engine and chassis remained identical to the RR 1200.

Buell, Specialists in a different kind of Harley-Davidson

On February 19, 1998, Erik Buell announced that he had sold almost all of the shares of Buell Motorcycle Company to Harley-Davidson, Inc., which had already been an active minority partner of Buell for the past five years. Erik Buell was named chairman and chief technical officer, while Jerry Wilke was offered the position of president of the new organization.

During the press conference Harley-Davidson held at Daytona Beach in March 1998, Erik Buell emphasized that Harley-Davidson's acqusition of the majority of the shares of Buell demonstrated the willingness of that firm to commit itself to the development of Buell as an independent entity:

"Now, thanks to increased support from Harley-Davidson, we have the assets we need to create the growth I have always dreamed of. The aquisition is a demonstration of confidence on the part of Harley-Davidson, both in myself and in the whole Buell team. And even though our business is already growing strongly, I can guarantee you that you haven't seen anything yet!"

In 1999, the inauguration of the research and development center, over 64,000 sq. ft (6000 sq. m) near the East Troy, Wisconsin factory, assured the modern and in-house development of Buell motorcycles. For Erik Buell, "It is there that the future of Buell is being developed" and the first results were soon visible; in 2000 the company stunned the motorcycle world with its introduction of a small, low motorcycle that was easy to handle in cities. So, after having produced only two-cylinder motorcycles, Buell produced its first single-cylinder bike in 2000. This motorcycle was the 492 cc Buell Blast, which was intended for an entirely new market in the United States (it was limited to the US market), where it quickly received acclaim for its innovative design. Constructed using a Sportster engine, from which one cylinder was removed, it featured reduced weight and an ease of use that made it an ideal vehicle for beginning motorcyclists as well as for short daily trips.

However, the main focus of Buell remained sportbikes, as superbly demonstrated by the XB9R Firebolt, a machine that renewed the sportbike concept with its unique form when it entered the Buell line in 2002. With its coded futuristic name, the Buell XB9R promised to bring innovations, and succeeded. It still used the traditional Sportster-type Harley-Davidson V-Twin engine, but one that delivered 92 hp at 7500 rpm and had a torque of 9.37 mkg at 5500 rpm. The mechanics of these Buells were not out-dated as the motor was entirely redesigned with a bore x stroke of 88.9 x 79.38 mm, numbers that appoach an over-square engine rather than the traditional long stroke Harley-Davidson engines. This Buell no longer used the old 1203 cc engine of the other models; instead, it made use of a 984 cc engine with electronic fuel injection and with the swing arm attached to the engine case, true innovations that made the XB9R a modern bike that still remained faithful to its traditional V-Twin origins.

Esthetically, this XB9R radically broke from the other Buells, as the large air filter vanishes from the left side of the bike and the frame becomes a visible aluminum twin-spar. The frame is particularly interesting because it is also used as a gas tank. The fuel circulates freely in the two runners and around the steering column, which has the advantage of lowering the center of gravity and freeing up the space necessary for the air box with a 45-mm diameter intake. But the innovations for this chassis are just not limited to these elements; the XB9R also has a larger exhaust compared to the other Buells because the rear shock absorber is no longer positioned below the engine. This modification was made possible due to other innovations regarding the frame. Specifically, the Buell V-Twin Sportster is a dry-sump engine, which means that a separate oil tank contains the lubricant. Previously, the oil tank was located beneath the saddle, but with the XB9R the swing arm functions as the oil tank. The freeing-up of the space where the oil tank was previously located allowed for placement of the rear suspension in a more conventional position, the cantilever-type found on the majority of modern sportbikes.

Analyzing the XB9R in more detail, other unusual innovations become apparent. The front braking system uses a simple perimeter floating brake disc and a six-pot opposed piston caliper, and the final drive is obtained by a toothed belt. This was a traditional solution, and one valued by Buell, but it was perfected by the adoption of a tensioner pulley. For this motorcycle, pure power with maximum hp and rpms was not the desired end. Instead, the focus was on experience and performance at every speed in order to provide true driving pleasure to the sportbike rider who wants to enjoy the XB9R in all conditions, while at the same time owning a unique motorcycle.

In 2003, the Buell line was completely changed and included just the Buell Blast (for the American market), the Firebolt XB9R and the Lightning XB9S. This new Lightning XB9S used the same chassis and engine as the Firebolt XB9R sportbike, but transformed it into a roadster by rearranging the parts and changing the driving position. It abandoned clip-ons in favor of a higher handlebar, and replaced the triple tree with a windshield; however, only the rear part of the motorcycle breaks entirely from the XB9R, in that the rear curve of the chassis now functions as the covering and supports a saddle lowered by 10 mm (765 mm), while the footrests are placed 25 mm lower.

340 The Buell Blast, powered by a single cylinder engine derived from a Sportster engine, is a small and lightweight motorcycle designed for beginners, which features modern technology and a frame that functions as an oil tank.

341 top An original concept, the new Buell 2002 XB9R chassis was modified to include 3a fake gas tank that hides the air box, while the fuel was contained in the frame and the oil in the swing arm.

341 center The Buell 2003 line was entirely redesigned and the Lightning XB9S, derived from the XB9R, was the only Roadster model that remained available that year.

341 bottom The technology of the Firebolt XB9R cleaned the slate by eliminating, in 2003, all of the old V-Twin Buells and announcing the future arrival of motorcycles built around this new chassis.

From Buell Harley-Davidson to Buell Rotax

While the Harley-Davidson Motor Company celebrated its centennial anniversary in Milwaukee at the end of August 2003, at the end of July of the same year Buell Motorcycle Company celebrated its 20th year of existence with a large gathering near its own factory in East Troy, Wisconsin, under the patronage of its parent company Harley-Davidson. Erik Buell used this opportunity to announce the recent transfer to Harley-Davidson of the remaining 2% that he held in Buell Motorcycle Company, which expanded its line with three new models for 2004. The Buell Firebolt XB12R and Lightning XB12S, that use the design of the XB9R and XB9S models, are driven by 1203 cc XB1200 engines which produce 103 hp and benefit from an air- and oil-cooling system; the Low XB9S is a low version of the XB9S.

Two new Buells were introduced in 2005: the Lightning CityX XB9SX (Erik Buell asks that it be pronounced as "Lightning City Cross") an evolution and replacement of the Lightning XB9S, and the Lightning XB12Scg, a low version of the XB12S, thanks to a lowered front and rear suspension and a narrower saddle. The Buell line comprised six models including the Blast, a diversification that indicated the good health of the company, which continued to have steady increases in the sale of V-Twins. The Lightning CityX is an evolution of the Lightning XB9S that has the advantage of being offered at a reduced price. This motorcycle reinforces its use as an urban bike with the name "City Cross," and surprises with its stylistic changes, such as the windshield and the fake gas tank in transparent blue plastic that adds color to the motorcycle while leaving some of the mechanical elements visible. Erik Buell, Chairman and Chief Technical Officer of Buell Motorcycle Company, presented his most recent model as ". . . a bike for the rider who loves to dodge traffic, power away from lights and utilize the incredible agility of a sport motorcycle to charge through the asphalt jungle."

While in 2006 Buell enhanced the line by adding the Lightning Long XB12Ss and the Ulysses XB12X, for a total of eight bikes, improvements were also made to all of the Buell V-Twins. The new swing arm, more rigid and lighter by 0.55 lbs (0.25 kg) with respect to the preceding model, still functions as the oil tank, but the cap was repositioned and redesigned to facilitate its access and opening. A new transmission allowed for a more rapid shifting with helical gears that are quieter and smoother than the norm. The air intake was relocated and integrated into the bodywork that encloses the fuel cap, allowing for an increased amount of fuel to be held in the chassis.

The Buell Lightning Long XB12Ss was a Lightning XB12S with an extended swing arm and a wider angle front fork, transformimg it into a bike with a more conventional character, and which was easier to handle thanks to improved ergonomics for the rider.

The Buell Ulysses inaugurated a new family, the Adventure, and was classified among the large street trails on the motorcycle market. It also preserved the unique appearance of the Buell personality since the introduction of the new perimiter frame. This model features an original accessory, the "triple tail" that covers the rear of the saddle and is able to pivot on its own axis for use either as a passenger backrest or a luggage rack.

While Buell produced its 100,000th motorcycle

342 top The XB12R, which was introduced in 2004, was a Buell that continued the technology of the XB9R chassis, but with an engine no larger than 984 cc rather than 1200 cc.

342 bottom While the XB12S was the only 1200 cc Buell Roadster version in 2002, in the following year the XB12Scg with a low seat was introduced.

342-343 The lightning CityX (City Cross) XB9SX was a 2005 evolution of the Buell XB9S intended for effective and enjoyable urban use, to dodge traffic and power away from lights.

343 top All of the Buells are assembled in the East Troy, Wisconsin Buell factory, to the southeast of Milwaukee.

in November of 2006, a Ulysses XB12X, the 2007 line was increased to include a new model, the Lightning Super TT XB12STT, bringing the total to nine models. This strategic addition is a derivative of the XB12 rather than a true innovation, and is placed between the Ulysses and the XB12S, with styling inspired by Supermotard motorcycles. Buell celebrated its 25th anniversary in 2008, with incredible innovations because Buell marketed a racing bike in 2007, the XBRR, produced in a limited edition of 50 bikes which enabled it to compete in Superbike competitions. The engine, taken from the XB12R Thunderstorm, but with an increased capacity of 1339 cc that delivers 150 hp to the crankshaft, is mounted to a modified XB12R chassis, to create a bike featuring true racing possiblities. However, even more suprising was Buell's unveiling of a bike that was not powered by a Harley-Davidson V-Twin engine.

344 top *The 2007 XBRR, a motorcycle whose use is limited to competitions, features an improved XB12R Thunderstorm engine whose capacity was increased to 1339 cc and provides 152 hp.*

2008: Buell abandons the Sportster engine?

Buell celebrated its 25th anniversary organizing various events, while the 2008 line underwent some interesting changes: the introduction of a model that launches a new generation of Buells no longer powered by Sportster-based engines.

The XB9R Firebolt was discontinued and the Buell line for 2008 was composed of 9 motorcycles: the Blast, the Firebolt XB12R, the 1125R, the Lightning CityX XB9SX, the Lightning XB12S, the Lightning Low XB12Scg, the Lightning Long XB12Ss, the Lightning Super TT XB12STT and the Ulysses XB12X.

As was logical, the 2008 XB line with V-Twin Buell engines enjoys an evolution based on the XBRR Buell racing bike engine from the previous year. Therefore, the 984 cc and 1203 cc air/oil-cooled Thunderstorm

engines are equipped with a new electronically synchronized Engine Control Module (ECM) that regulates the fuel injection system, new crank pins and an improved oiling system. These modifications required a redesign of the engine case; now having a catalyzed exhaust system, this Buell meets Euro 3 regulations. However, the main innovation was the introduction of the 1125R as a new platform, replacing the XB9R and thereby eliminating engines derived from the Harley-Davidson Sportster, in contrast to all the XB models.

The Buell 1125R is a truly innovative sportbike powered by a V-Twin 146 hp engine, and with a dry weight of 375 lb (170 kg), that has become the company's new high-end sport road bike for the year 2008. The motorcycle maintains the spirit of other Buell creations, with a short frame that also acts as a gas tank and on which is mounted the engine christened Helicon, a 72° angle water-cooled V-Twin that was created through a partnership between BRP (Bombardier Group), Rotax, and Buell. This new engine, assembled by BRP-Rotax in Austria is made exclusively for Buell, and the 1125R is assembled at the Buell factory in East Troy, Wisconsin.

Buell motorcycles with Sportster engines still exist and will continue to exist in 2008, but in the future, if a line based on the new Helicon motor is successful, what will happen to Buells powered by modified Sportster engines?

But, like the other manufacturers, Harley-Davidson is hit by the trade depression and his profit drops. At the end of September 2009, the Company posted a profit of $26.48 million, an 84% retreat from $166.54 million, or 71 cents a share in the year-ago period. The latest quarter included a one-time charge of $14.2 million related to Buell...

October 15 of 2009: after 26 years spent building American sportbikes, Erik Buell's eponymous motorcycle company has reached the end of the road. Harley-Davidson announces that it will discontinue the Buell product line. Harley-Davidson CEO Keith Wandell said: "We must focus both our effort and our investment on the Harley-Davidson brand, as we believe this provides an optimal path to sustained, meaningful long-term growth".

344-345 The Buell 2008 line offered a true innovation with the 1125R (first motorcycle on the left) powered by an engine that was no longer derived from a Sportster, which is produced by Rotax and Bombardier in collaboration with Buell, exclusively for this bike. This is the only true sports model in a line comprising nine motorcycles with a predominate Roadster character.

346 The Lightning XB12Ss, which was introduced in 2006, is an evolution of the XB12S with the wheel base increased by 40 mm to 1360 mm, a front fork angle increased from 21 to 23.5° and a trail that increased from 84 mm to 119 mm.

347 The Buell Ulysses XB12X inaugurated in 2006 a new family, the Adventure. This motorcycle featured unusual equipment, like a protective grill for the headlights and a windshield in the form of a spoiler.

Harley-Davidson VR 1000
1988 - 2001

The VR 1000 project began in 1988 when the first sketches of the engine were drawn and then entrusted to Roush, a company specializing in high performance and Nascar automobile parts, to finalize the technical study in 1989. A prototype of the engine was created, then Harley-Davidson integrated the project into the "Harley-Davidson Racing and Engineering" program in the hands of Steve Scheibe, a talented young engineer from Roush. This engineer, who had already been working on the VR 1000 project, primarily on the cylinder heads and the fuel injection, joined the Milwaukee company in the early 1990s.

The resulting engine was a 60° V-Twin with fuel injection, twin overhead cams on each cylinder, liquid cooling and cases with horizontal couplings. It delivered 140 hp and a maximum engine speed of 10,850 rpm, to power a motorcycle weighing 160 lbs (73 kg) dry. The frame was produced under the direction of Mike Eatough, who arrived at Harley-Davidson when the company Armstrong was aquired by Harley-Davidson in 1986. This company produced an off-road military motorcycle with a single cylinder Rotax engine, that was thereafter marketed by Harley-Davidson.

On paper, the VR 1000 motorcylce was an excellent racing machine, having the necessary specifications to beat Ducatis, which dominated the Superbike category in the early 1990s thanks to the 851 and 888 models with similar mechanical configurations. At this time, the Japanese manufacturers who had abandoned this type of competition were gradually returning. Unfortunately the VR 1000 participated in its first race, the Daytona 200 only in 1994, and the test revealed that the motorcycle was not yet ready for a long race of this type, suffering from both electrical and mechanical problems. It was not until 1996 that the season went smoothly, with some good results, but the motorcycle was no longer competitive against its more modern rivals. The competition budget for the VR 1000 was re-evaluated every year until 2001, when the introduction of the V-Rod hastened the end of the American Superbike, and its competition budget was eliminated for 2002. After eight seasons of American Superbike racing, the VR 1000 results were a pole position and a few podiums but no victories.

In order to be approved for the American Superbike championships, the VR 1000 was produced in a series of 50 at the astonomical price of 400,000 FF at the time. Competition turned out to be the best possible laboratory for the development of the motorcycles, and the VR 1000 allowed Harley-Davidson to develop new technologies, in particular fuel injection and liquid cooling, which were used on the new V-Rod that was introduced in the summer of 2001 and christened VRSC (V-Twin Racing Street Custom). When the Harley-Davidson Motor Factory announced the halt of the VR 1000 racing program, they pointed out that they were not giving up racing, as they would continue to participate in dirt track races and they unveiled an ambitious Dragster program based on the V-Rod.

***348 top** The engine of the VR 1000, a 60° V, provides 140 hp at a maximum speed of 10,850 rpm and has a weight of 160 lbs (73 kg).*

***348 bottom** The VR 1000, which participated in its first race in 1994, was a particularly advanced, high-performance racing machine at the time it was initially designed, but by the time it was finished, it could not effectively compete against the bikes of the other manufacturers.*

Dragster V-Rod
2002

The Motor Factory entered an elaborated version of the V-Rod in the NHRA (National Hot Rod Association) Dragster championship in the Pro Stock category, thus creating the team "Screamin' Eagle/Vance & Hines" in 2002. During the first season, dedicated to fine tuning and development, the V-Rod Screamin' Eagle piloted by GT Tonglet showed good potential, but did not qualify for any races. In its second season (2003), however, another pilot named Andrew Hines joined the team and qualified for all 15 of the NHRA championship races in the Pro Stock Motorcycle category, participating in the final race for two of these.

Andrew Hines then won the Pro Stock 2004 championship and became, at the age of 21, the youngest champion in the history of the NHRA. It was the first title for Harley-Davidson in Dragster, and also the first victory for a Twin liquid against a line-up of motorcycles powered by four cylinder engines.

Harley-Davidson won the Pro Stock title for the second consecutive year in 2005, again with Andrew Hines, and GT Tonglet his Screamin' Eagle teamate placed second in the championship. The champion once again for the 2006 season with the V-Rod Screamin' Eagle, Andrew Hines once again entered the record books as the first pilot to win three consecutive championships and as the record holder for six of the top ten recorded speeds. These results comfirmed the excellent potential available when starting with a V-Rod as the mechanical basis, and anticipates a future rich in variations of the V-Rod and its Revolution engine.

***349 top** The VR 1000, painted black on the left side and orange on the right to symbolize the Harley-Davidson Racing Team colors, continued its racing career until 2001.*

***349 bottom** The official Harley-Davidson team, the Screamin' Eagle/Vance & Hines Team, participated in Dragster competitions starting in 2003 with a motorcycle based on a V-Rod, and went on to win the 2004 championship.*

BUELL: XB12STT LIGHTNING SUPER TT

Erik Buell stated: "The Buell Lightning Super TT XB12STT is a high performance motorcyle with an urban character!"

In fact, like all of the Buell V-Twins of 2003 this XB12STT continues the basic concpept of the XB that was introduced in July 2001, and was integrated into the Buell line in 2002. While the lateral licence plates are an innovation, the small windshield, the lights protected by grilles, the handlebar and the hand grips come from the XB9 City Cross, while the XB 12STT is an XB12Ss with higher clearance suspension, but remaining lower than the Ulysses, from which it takes its front fender. This Buell combines technologies from various models of the brand, and has styling inspired by Supermotard motorcycles. However, it should not be forgotten that the STT, with its high saddle, is powered by a large 1200 cc V-Twin and therefore has a considerable weight. This is a fact that should be kept in mind during the excitement of the action.

The long stroke V-Twin (of Harley-Davidson Sportster origins) with air and oil cooling, and a balanced distribution, has benefitted from a thorough study by Buell to increase the power and to offer a high level of performance and enjoyment. The engine, with electronic fuel injection and equipped with an exhaust system featuring valves, delivers over 100 hp and has a generous torque of 11 da.Nm at 6000 rpm. These are data comparable to the specifications of the competing over-square type engines of the other brands, which have more power but usually provide less enjoyment. Having a piston stroke greater than their diameter, with a bore of 88.9 mm and a stroke of 96.8 mm, torque is favored over power. However, this long stroke engine succeeds in offering good performance at any engine speed and therefore combines incredible driving enjoyment with pure excitement. The engine of the XB12STT, sufficiently light enough to enable high performance even at the lowest engine speeds, is not limited to recreational use, contrary to what might be assumed from its styling. The motorcycle reveals itself to be versatile for touring and longer rides, regardless of the pace, as the engine eagerly responds with minimal prodding, under any circumstances.

Curiously, the driver of this Buell is situated toward the front, in a driving position that is disconcerting. However, once the ergonomics are learned, and with a frame that quickly reveals its rigidity and optimal handling, an exciting and enjoyable ride is provided while control of the motorcycle is perfected. The engine is steady without being rough even at the lowest of speeds, and welcomes the driver to take full advantage of the motorcycle thanks to its general ease of use. In fact Buell unveiled the XB12STT as a versatile and enjoyable motorcycle havinga sportbike character that is easy to make the most of and whose use comes closer to that of a more conventional Harley-Davidson.

***350 bottom** Representing a new generation of the classic Buell with a Sportster derived engine, the XB12STT Lightning combined the elements of various Buells to create a motorcycle with behavior and styling that was half-Roadster and half-Supermotard.*

***350-351** From the appearance of the first V-Rod models, both neophytes and specialists used them as a base for creating a personalized motorcycle, with a resized rear wheel, as Harley-Davidson itself offered standard several years later.*

Harley-Davidson Revolution engine 1250

V-ROD: VRSCDX NIGHT ROD SPECIAL

In the summer of 2001 Harley-Davidson introduced the new V-Rod, and the company spokesperson proclaimed: "It is time to challenge the senses and the status quo. The newest offering from Harley-Davidson is poised to trounce all previous conceptions of not only a Harley-Davidson custom, but custom motorcycles in general.

The 2002 Harley-Davidson line unveiled this new machine that broke with tradition in its use of decidedly modern technology. With this V-Rod, the Milwaukee manufacturer maintained the spirit of the brand and also succeeded in attracting the attention of many who resisted the idea of purchasing a Harley-Davidson. The stylish design of this motorcycle, a Dragster type but a little less aggressive, is not unrelated to many motorcyclists' change of opinion about such a purchase.

The V-Rod comfortably holds the rider with a very low saddle (only 660 mm high), even though the handlebar and the footrests are somewhat far away for smaller drivers. The problem is easily resolved, as Harley-Davidson offered an optional saddle equipped with a backrest that was moved forward 1 inch (2.5 cm), and footrests that could be raised higher by 1 inch (2.5 cm). Once on the street, the V-Rod, with its two simultaneously moving pistons, offers a true Harley-Davidson sound, but with almost no engine vibrations transmitted to the driver. The engine provides a full sensation with good torque and powerful acceleration. The lightweight Revolution reaches 2000 rpm in 5th gear, then runs enthusiastically starting at 4000 rpm and pulls strongly when passing 6000 rpm. The increase in the hp is obvious, with a motorcycle that displays a Dragster-type behavior. This engine launched a new family, alongside the Sportster, Dyna, Softail and Touring. With interesting development possibilities, it quickly diversified into various models targeted for specific customers. In 2008 the engine gained a few more cubic centimeters and the V-Twin Revolution changes from an 1130 cc to a 1250 cc.

The VRSCDX Night Rod Special Harley-Davidson was created to meet the requirements of those desiring to purchase and then customize a Night Rod. While the majority of motorcylces claim versatility, comfort, and ease of use, the Night Rod Special has an exclusive character that is destined only for expert motorcyclists.

With respect to the first generation V-Rods, the Night Rod Special is powered by a redesigned engine mounted on a modified chassis and offers additional features. The old frame could not incorporate the large 240-mm wide tire that replaced a 180-mm tire, or the new 19-liter gas tank (previously 14-liter) which provides increased independence. This VRSCDX remains a recreational motorcycle, for the enjoyment of short runs and urban use, since the driving position does not really encourage long trips.

The pilot, boxed in behind the large gas tank, with arms and legs straining forward to grip the small handlebar and the controls, gets a workout with every ride. While long journeys seem unlikely, the pleasure of the Night Rod is found elsewhere. Seated low, with an engine that growls through the new exhaust system, the driver is immediately inspired to open the throttle. The radical driving position is then forgotten in the pleasure of acceleration, playing with the engine's power in the intermediate gears, experiencing the perfect braking, and just delighting in its impeccable road-hugging performance. The massive rear tire demands that the motorcycle be forced when entering curves. It is a technique to be learned and then enjoyed, entering a curve, appreciating the bike's stability and rigidity, and then exiting the curve at a higher speed. This motorcycle is meant to be driven hard, as can be surmised from the design, and hard driving maximizes its character and effectiveness for road or city use. It is a true street Dragster, practically without compromise, and not simply a Custom with an aggressive design and toned-down behavior, because if the V-Rod is THE tough motorcycle par excellence, the VRSCDX represents its intense, concentrate version.

352 The Night Rod Special VRSCDX was a new V-Rod with a modified frame, a 240 mm-wide rear tire and a 1250 cc engine that replaced the 1130 cc engines of the first generation V-Rods.

352-353 While Harley-Davidson already offered beautiful, highly stylized Custom models, it opened the Touring motorcycle market with this Street Glide, where performance and aesthetics combine with comfort.

Harley-Davidson Twin Cam 96 engine

ELECTRA GLIDE: FLHX STREET GLIDE

The FLHX Street Glide, which was essentially based on the well known Harley-Davidson Touring, as are both the Electra Glide and the Road King, is powered by the 2007 Twin Cam 96 engine with fuel injection and mounted on Silentbloc to absorb vibrations, as are all the Dyna and Electra models. The major difference in the Street Glide is the lowered rear part and the reduced equipment. In short, it is either an Electra Glide Classic in a lightened form and simply customized, or an Electra Glide Standard that is better equipped and slightly customized. The Street Glide offers comfortable ergonomics for the driver, with an accommodating seat and with feet placed forward. Nevertheless, the Custom configuration makes steering a little more sluggish than the Electra Glide Standard, with the FLHX being slightly less agile and not as easy to handle due in part to its unique design. This behavior, though, is easily assimilated and compensated for by the pleasure derived from gliding on this beautiful American bike that has the best established mechanical temperament of any 2007 bike.

In comparison with the old Street Glide powered by the 1450 cc Twin Cam 88, the Street Glide powered by the 1584 cc Twin Cam 96, has more decisive pick-up and acceleration in the lower gears, making one forget the lack of performance of the 2006 model at low speeds. The Twin Cam 96 allows acceleration in the higher gears without having to play with the shifting, as often happened before. While the Twin Cam 88 was not particularly enthusiastic below 3000 rpm, and only truly expressed itself above that, the Twin Cam 96 runs well starting from 1500 rpm, growing steadily up to 4000 rpm and maintains steady delivery up to 5500 rpm. This 1584 cc performs better than its predecessor at all speeds. At equal velocities, its speed of rotation is lower than that of the 1450 cc thanks to the different distribution of the new 6-speed transmission, which reduces gas consumption and lengthens the engine lifespan.

The FLHX, like the other Big-Twin models, enjoys these benefits and succeeds in merging the spirit of a Harley-Davidson road bike with that of a Harley-Davidson Custom, to stand out as an enjoyable, relatively versatile motorcycle that also attracts attention.

DYNA GLIDE: FXDF FAT BOB

The 2008 Dyna Glide Fat Bob, the only motorcycle in the Dyna family that was not updated from a Twin Cam 88 to a Twin Cam 96 engine, combines features characteristic of the Dynas with a different design and a new personality. The name Fat Bob refers to the shape of the gas tank that has been the trademark of custom Harley-Davidsons for many years. Powered by a Twin Cam 96 engine that supplies 12.9 da. Nm of torque and a 6-speed Cruise Drive transmission, the Fat Bob features a unique air filter and a new exhaust system called the "Tommy Gun" with a 2-1-2 collector as on the V-Rod. The drag bar handlebars hold the electrical wires, demonstrating the particular care given to the styling, which is also apparent in the double headlight placed between the polished front fork tubes and the wide front and rear tires mounted on the 16-inch (406-mm) slotted aluminum rims.

This motorcycle combines the street attitude of the Dyna Glide with the spirit of the Softail Custom, but has a wheel base and a lean angle that are less pronounced than those of the Softail Rocker. While the pleasure of the road prevails with the Chopper styling, comfort is not forgotten thanks to a firm rear suspension and the easily manageable clutch and gear shift. Sixth gear is used primarily for road use and the new 1584 cc Twin Cam 96 reveals a satisfying road attitude. The engine runs smoothly at any speed and is even more powerful than the FLHX Street Glide with vigorous performance from the first acceleration. With the handlebars and the driving position similar to that of a Dragster, it is very tempting to open the throttle of the Dyna Glide Fat Bob at every acceleration, while the Street Glide invites a more touring attitude. Since this Twin Cam 96 accelerates and responds to the slightest encouragement, while maintaining great performance, it is difficult to resist this pleasure. Until recently, the Dyna Glides were known for their road character and their restraint, slowly revealing their strong points over the miles, but now these models have been retailored for a market that demands immediate pleasure.

SPORTSTER: XL 1200 NIGHTSTER

The retro-style Nightster is a Sportster reduced to its essentials, creating a minimalist effect both sober and elegant, which is also suitable for its intended use – the pleasure of brute force and immediate response.

With its single seat, spoked wheels, two shock absorbers and the "peanut" gas tank, this retro-inspired Sportster is reminiscent of motorcycles that were stripped and modified in the 1960s. No uneccessary element disturbs the well-considered sobriety, but the minimalist accessories do provide all of the necessary comforts of a modern motorcycle. The single seat, the small gas tank, and the pervasive black color visually reduce the volume of the Nightster, which appears to gather in on itself. The low seat and the small size reassure and invite comfortable control, despite a dry weight of 545 lbs (251 kg) that is forgotten once in the saddle. The feeling of simplicity and ease of the commands allow for an maneuverable motorcycle, despite the expressive roar of the engine that creates excitement right from the start. The credit goes entirely to the engine that recreates the pure pleasure of the old XLX. Riveting, the engine quickly seduces when driving the motorcycle, thanks to a hidden readiness that is instantly awakened. With a high torque even at low speeds, the engine always delivers regardless of the shape of the roads. It is enough to accelerate out of the turns and cut the gas when approaching the next one to cover the miles quickly and enjoyably. Despite the long stroke configuration, the V-Twin offers an incredible smoothness and a steady delivery during acceleration, accompanied by the distinctive, captivating Harley-Davidson sound. Since the weight-to-power ratio is very favorable in comparison to the Harley-Davidson Big-Twins, it has suprising acceleration and a recreational feel that are perfectly adapted to urban or suburban use, as well as trips on mountain and country roads. The impeccable handling due to its balanced equilibrium, the reduced wheel base and the handlebar desinged to create suprising responsiveness make this an extremely effective motorcylce for these routes. The stability is not diminished and the effectiveness is even improved, except on poorly paved roads, thus creating a Sportster that has been particularly well furnished with the profile of an "underground" vehicle destined for pure pleasure. In fact, this Nightster's possibilities have no limits.

***354 top** The FXDF Fat Bob, which has abundant torque and power thanks to its Twin Cam 96 engine, offers the immediate pleasure of powerful acceleration and pick-up in a bike with an amazing street character.*

***354-355** The 2008 Sportster XL 1200N Nightster completes the European line. It has a greater height from the ground with respect to the 2007 and 2008 American models, but maintains the light system that combines the stop and directional signals, preserving the lines of this motorcycle dedicated to pure driving pleasure.*

Buffalo Billiards
Wild
201
Service Entrance

Harley-Davidson Twin Cam 96B engine

SOFTAIL: FXCW ROCKER ET FXCWC ROCKER C

The Softail family is powered by the Twin Cam 96B version that features internal counter-balancers which reduce vibrations. The engine is mounted directly on the rigid frame, in contrast to the other Big-Twins that mount the Twin Cam with Silentbloc.

These two Softail Rockers use a Softail frame system, with the rear part of the motorcycle imitating the profile of the famous Hardtail rigid frame, but with the suspension hidden under the engine and the transmission. The Rocker accentuates its Chopper lines with the stylish rear fender in metal, which attaches to the swing arm without visible support, and covers the large 240-mm wide rear tire. It has a pronounced steering column be transformed into a dual saddle in a few seconds, thanks to an ingenious system that unfolds from the driver's seat.

The 96B version of the Twin Cam engine has the smoothness and the acceleration of the Twin Cam 96 with decisive acceleration at less than 1500 rpm and pulls strongly above this. The pick-up and the acceleration are good at any speed, and perfectly match the expectations for a custom motorcycle with such radical styling. The handling, even with the large rear tire and the considerable weight of the motorcycle, is suprisingly good thanks to the equilibrium provided by the low center of gravity. The motorcycle handles well, particularly when riding fast through twisted streets, where it asserts itself with a simple turn of the throttle

angle of 36.5° and a very low seat at 622 mm from the ground on the Rocker standard. The resulting combination has a flowing style, which is continued by the double handlebar holding a large part of the wiring. Technically similar, in terms of both the frame as well as the engine, these two motorcycles are distinguished by their finishing and their saddles. While the standard version has a single seat, the Rocker C has a seat that can coming out of the curves. Although the wheel base is long, the motorcycle is equilibrated and stable, with a front end that is easy to steer and a back end that assures power. By introducing these two motorcycles, each having a distinctive styling and unusual technology for motorcycles produced in series, Harley-Davidson steps foward as the first to manufacture Custom Chopper-style motorcycles not just as a limited edition but as a part of the line.

356 The Softail FXCWC Rocker is different from the Rocker Standard for its styling, and more importantly for its ability to carry a passenger thanks to the system located beneath the driver's seat, which unfolds to make a passenger seat.

357 The Softail Rocker and Rocker C, which were introduced in the 2008 line, have a decided Chopper look thanks to the rear fender that attaches to the swing arm and the standard low and long lines.

Harley-Davidson Models

Beginning in 2007, the Harley-Davidson Big-Twins were powered by a new engine, the 1584 cc Twin Cam 96; the Sportster engines used electronic fuel injection as did the other Harley-Davidson models; and the V-Rods were all powered by a Revolution engine whose capacity was increased from 1130 cc to 1250 cc in 2008. The 2008 Harley-Davidson line is composed of three groups of motorcycles powered by different engines. Two of these groups, the Sportsters with the transmission inside the engine case, and the Big-Twins with the transmission separate from the engine case, use traditional air cooling, while the V-Rod engines are liquid-cooled. These three groups are subdivided into families that include different types of street, custom, retro and touring motorcycles. The main group, which uses the 1584 cc Big-Twin engine, is divided into families based on specific styles that are destined for various uses.

The Dyna Glide family, with a Twin Cam 96 engine, a traditional frame and two shocks conventionally placed to the rear, offers a street motorcycle for agile driving.

The Touring Electra Glide family, with a Twin Cam 96 engine and a conventional frame, is equipped for touring, with saddle bags, bodywork and windshield, has less of a sport feel than the Dyna Glide, but provides more comfort and protection.

The Softail family, with the Twin Cam 96B engine with anti-vibration counter-balancers that is mounted without Silentbloc on a rigid-style frame, evokes the image of traditional Harley-Davidsons, Customs and even Choppers.

The Sportster group still uses the 883 cc and 1200 cc V-Twin Evolution engines, and is divided into roadster, sport and custom models, that are lighter, more lively and more accessible motorcycles than the Big-Twins.

The third group is the V-Rod, with Revolution engines, which are clearly distinguished from the others, both for the liquid-cooled engine and for the Dragster styling. The models that comprise this group all assert the same fast, high-strung spirit that enjoys being pushed, as their design shows. With a customized appearance from the start, they attract a new clientele that demands high performance.

The Dyna family is composed of the FXD Super Glide, the FXDC Super Glide Custom, the FXDB Street Bob, the FXDL Low Rider, the FXDF Fat Bob, and the FXDWG Wide Glide.

The Touring family is composed of the FLHR Road King, the FLHRC Road King Classic, the FLHX Street Glide, the FLTR Road Glide, the FLHT Electra Glide Standard, the FLHTC Electra Glide Classic, and the FLHTCU Ultra Classic Electra Glide.

The Softail family contains the FXSTB Night Train, the FXSTC Custom, the FLSTF Fat Boy, the FLSTN Deluxe, the FLSTC Heritage Classic, the FXCW Rocker, and the FXCWC Rocker C.

The Sportster family is divided into the 883 cc models and the 1200 cc models, with the XL 883, the XL 883L Low, the XL 883C Custom, the XL 883R, the XL 1200N Nightster, the XL 1200C Custom, the XL 1200R Roadster, and the XL 1200 L Low.

The VRSC V-Rods are the VRSCAW V-Rod, the VRSD Night Rod, and the VRSCDX Night Rod Special.

Harley-Davidson Nomenclature

FL/Electra Glide: Full dresser, low compression, 4-speed, engine mounted without Silentbloc.

FLH/Electra Glide: Full dresser, high compression, 4-speed, engine mounted without Silentbloc.

FLF/Electra Glide: Model with foot shift.

FLHS/Sport Solo: Electra without bodywork, 4- and 5-speed.

FLHT/Electra Glide: Full dresser, engine mounted with Silentbloc, 5-speed, bodywork on the fork

FLHT/FLHTI Electra Glide Standard: Based on an Electra Glide, but without the trunk and with a minimal instrument panel, for an attractive price, from 1996. 5- and 6-speed. Available with carburetor or fuel injection from 2003. Fuel injection only from 2007.

FLT/Tour Glide: Same, but with bodywork on the frame

FLTC/Electra/Tour: Same, with two-tone paint

FLHTC/FLHTCI/Glide Classic: Tour Pack, optional radio/cassette. 5- and 6-speed. Available with carburetor or fuel injection from 1996. Fuel injection only from 2007.

Electra Glide Standard FLHT

Electra Glide Classic FLHTC

Road King FLHR

Road King Classic FLHRC

Street Glide FLHX

Ultra Classic Electra Glide FLHTCU

FLHR/FLHRI/Road King: Lightweight Electra Glide, with removable windshield and saddlebags for a managable touring bike. 5- and 6-speed. Available with carburetor or fuel injection from 1996. Fuel injection only from 2007

FLHRCI/Road King Classic: A Road King with fuel injection, spoked wheels and leather saddlebags for a vintage look. 5- and 6-speed.

FLHRS/FLHRSI/Road King Custom: A Road King without a windshield, with saddlebags and a low and heavy line. 5-speed. Available with carburetor or fuel injection from 2004.

FLHX/Street Glide: An Electra Glide Standard, with Electra Glide Classic finishing and a low line. 6-speed with fuel injection.

FLHTCU/FLHTCUI/Ultra Classic Electra Glide: Tour Pack, radio/cassette, electronic cruise control, CB, Intercom. 5- and 6-speed. Available with carburetor or fuel injection from 1996 and only with fuel injection from 1999.

FLTR/FLTRI/Road Glide: An Electra Glide Standard with bodywork attached to the steering column and an audio system. 5- and 6-speed. Available with carburetor or fuel injection. Fuel injection only from 2003.

FX/Super Glide: Smaller fenders, straighter fork, 4-speed.

FXE/Super Glide: Same, with electric start.

FXEF/Super Glide: Same, with double gas tank.

FXS/Low Rider: The first Low Rider (with chain in 1977).

FXB/Sturgis: Factory custom, belt drive (1980-1982).

FXSB/Low Rider: Same, but with single belt final drive.

FXWG/Wide Glide: Factory custom, wide front fork, 21-inch front wheel, Fat Bob fender.

FXEF/Low Rider: Electric start and Fat Bob gas tank.

FXFB/Low Rider: Low Rider basic with kick and Fat Bob.

FXDG/ Low Rider: Rear disc wheel (Disc Glide in 1982).

Dyna Low Rider FXDL

FXR/Super Glide: New frame, 5-speed, engine mounted on Silentbloc, without accessories (the name was originally Super Glide II, but the II quickly disappeared).

FXRS/Low Glide: Same, with diverse Low Rider options and from then on called Low Rider.

FXRS-Sp/Sport Edition: Same, with the suspension 2 inches higher.

FXRS-CONV/Low Rider: Same, with steering column and saddlebags.

FXLR/Low Rider Custom: Customized Low Rider with a leather tank cover and a rev counter on the handlebar.

FXRC/Low Rider: Low Rider Custom from 1985 with candy paint and chrome engine.

FXRT/Sport Glide: Low Rider with bodywork and touring bags in fiberglass.

Dyna Wide Glide FXDWG

Dyna Super Glide FXD

Super Glide Custom FXDC

Dyna Street Bob FXDB

FXRD/Sport Glide: Same, with radio and Grand Touring options.

FXDL/FXDLI/Dyna Low Rider: Low Rider with new frame, new engine mounting system with Silentbloc and the 5- and 6-speed transmission box covered by the oil tank (model preceded by the Dyna Sturgis and Dyna Daytona versions). Carburetor or fuel injection from 2004. Fuel injection only from 2006.

FXDWG/FXDWGI/Dyna Wide Glide: Same, with a wide fork. 5- and 6-speed. Carburetor or fuel injection from 2004. Fuel injection only from 2006.

FXD/FXDI/Dyna Super Glide: Super Glide based on a Dyna. 5- and 6-speed. Carburetor or fuel injection from 2004. Fuel injection only from 2006.

FXDS-CONV/Dyna Convertible: Low Rider based on a Dyna with steering column and saddlebags.

FXDX/FXDXI/Dyna Super Glide Sport: Dyna Super Glide with double front disc, low saddle and suspension with more clearance. Carburetor or fuel injection from 2004.

FXDXT/Dyna Super Glide T-Sport: Dyna Super Glide with steering column, removable nylon saddlebags, wide saddle and suspension adapted to fast touring.

FXDC/FXDCI/Super Glide Custom: A Super Glide based on a Dyna but with chrome, additional paint and accessories. 5- and 6-speed. Carburetor or fuel injection. Fuel injection only from 2006.

FXDBI/Dyna Street Bob: Based on a Dyna Super Glide with a retro look, medium ape-hanger handlebar. 6-speed. Fuel injection.

FXDI/Dyna 35th Anniversary Super Glide: Super Glide Dyna commemorative edition based on a Low Rider. 6-speed and fuel injection. Produced only for one year.

FXDF/Dyna Fat Bob: Factory custom based on a Dyna with Wide Glide fork, fat tires, two headlights and a Fat Bob gas tank. 6-speed. Fuel injection.

Dyna Fat Bob FXDF

FXST/Softail: Rigid frame look with suspension, 4- then 5-speed.

FXST/FXSTI/Softail Standard: New name when it re-enters the line in 1999 with 5- then 6-speed from 2007. With carburetor or fuel injection from 2001. Fuel injection only from 2007.

FXSTC/Softail Custom: Softail custom version. Rear disc wheel and chrome. 5-speed.

FXSTC/Softail Custom: Re-enters the line in 2007. 6-speed. Fuel injection.

FXSTS/FXSTSI/Softail Springer: Softail with Springer fork. 5-speed. With carburetor or fuel injection from 2001.

FLST/Softail Heritage: Softail Hydra Glide fork. 5-speed.

FLST/FLSTI/Heritage: Re-enters the line in 2006. 5-speed. Available with carburetor or fuel injection.

FLSTC/FLSTCI Heritage Classic: Softail Heritage with 1960s look. 5- and 6-speed. Available with carburetor or fuel injection from 2001. Fuel injection only from 2007.

FLSTF/FLSTFI Softail Fat Boy: Softail with disc wheels. 5- and 6-speed. Available with carburetor or fuel injection from 2001. Fuel injection only from 2007.

FLSTN/Heritage Nostalgia: Softail Heritage with calfskin finishing. 5-speed.

FXSTSB/Bad Boy: Softail Springer that combines vintage and modern styles, painted entirely in black. 5-speed.

FXSTB/FXSTBI/Night Train: Softail Custom (telescopic fork) painted entirely in black. 5- and 6-speed. Available with carburetor or fuel injection from 2001. Fuel injection only from 2007.

Softail Custom FXSTC

Softail Heritage Classic FLSTC

Softail Fat Boy FLSTF

Softail Deluxe FLSTN

Softail Night Train FXSTB

Softail Rocker FXCW

Softail Rocker C FXCWC

FXSTD/FXSTDI/Softail Deuce: Softail with narrow lines and a rear disc wheel. 5- and 6-speed. Available with carburetor or fuel injection. Fuel injection only from 2007.

FLSTS/FLSTSI/Heritage Springer: Softail Springer with Vintage design, inspired by a 1948 Panhead. 5-speed. Available with carburetor or fuel injection from 2001.

FLSTN/FLSTNI/Deluxe: Nostalgia type Softail but without calfskin. Whitewall tires, Fat Boy fork. 5- and 6-speed. Available with carburetor or fuel injection. Fuel injection only from 2007.

FLSTSC/FLSTSCI/Springer Classic: : Softail Springer with many elements in black that are reminiscent of the Harley-Davidson style of 1948. 5- and 6-speed. Available with carburetor or fuel injection. Fuel injection only from 2007.

FXCW/Softail Rocker: Factory custom-based on a Softail with fender attached to the swing arm close to the wide rear tire. 6-speed. Fuel injection.

FXCWC/Softail Rocker C: A Softail Rocker with an ingenious removable passenger seat.

Sportster XL 883

Sportster XL 883C Custom

Sportster XL 883R

Sportster XL 883L Low

XL/Sportster: The first Sportster from 1957 that was followed by all the other Sportsters. 4-speed. Carburetor.

XLCH/Sportster: The first high performance Sportster, kick starter. 4-speed. Carburetor.

XLH/Sportster: Same, with electric start, later the Evolution version reuses the name. 4-speed. Carburetor.

XLT/Sportster: Touring equipment. 4-speed. Carburetor.

XLCR/Café Racer: The first and only Cafe Racer HD.

XR 1000/Sportster: Sport version inspired by the XR 750 race bike.

XLS/Roadster: Sportster Custom version. 4-speed. Carburetor.

XLX/Sportster: Version entirely in black with the Sportster engine in cast iron, reduced price.

XLH 883/Sportster: Standard version. 4- and 5-speed. Carburetor.

XL 883/Sportster: Standard version. 5-speed. Carburetor. Fuel injection from 2007.

XLH-DLX 883/Sportster: Deluxe version. 4- and 5-speed. Carburetor.

XLH-HUG (then Hugger) 883/Sportster: SSportster low version. 4- and 5-speed. Carburetor.

XL 883C/XL 53 C/Sportster Custom 53: Sportster 883 cc custom with rear disc wheel. 5-speed. Carburetor. Fuel injection from 2007.

XL 883R/Sportster 883 Racing: Sportster with two into one exhaust, black engine and XR 750 type racing colors. 5-speed. Carburetor. Fuel injection from 2007.

Sportster XL 1200C Custom

Sportster XL 1200R Roadster

XL 883L/Sportster: Sportster based on a Low. 5-speed. Carburetor. Fuel injection from 2007.

XLH 1100 Sportster: Standard version. 4-speed. Carburetor.

XLH 1200/Sportster: Standard version. 4- and 5-speed. Carburetor.

XL 1200S/Sportster: Sportster 1200 Sport. 5-speed. Carburetor.

XL 1200C/Sportster: Sportster 1200 Custom. 5-speed. Carburetor. Fuel injection from 2007.

XL 1200R/Sportster: Sportster 1200 Roadster, 1200 base version. 5-speed. Carburetor. Fuel injection from 2007.

XL 1200L/Sportster: Sportster Low version. 5-speed. Carburetor. Fuel injection from 2007.

XL 1200N/Sportster: Sportster Nightster, a customized retro and stripped version. 5-speed. Fuel injection.

XL 50 50th Anniversary: 1200 Sportster 50th Anniversary commemorative edition. A Roadster with anniversary logo and styling. 5-speed. Fuel injection. Produced only for one year.

Sportster XL 1200L Low

Sportster XL 1200N Nightster

Night Rod VRSCD

VRSCA V-ROD/V-Rod: New Harley-Davidson with 60° V 1130 cc engine, double ACT, 4 valves per cylinder, liquid-cooled, fuel injection and low Custom line.

VRSCB V-ROD/V-Rod: Same, with less finishing, black frame and a lower price.

VRSCR/Street Rod: As with the VRSCA above with different exhaust, additional 5 hp and 3 discs.

VRSCD/Night Rod: As with the VRSCR above with frame finishing, black equipment and different finishingpieces. 1130 cc then 1250.

VRSDX/Night Rod Special: An entirely black Night Rod with 240-mm wide rear tire. 1130 cc then 1250 cc.

VRSCAW/V-Rod: Replaces the VRSCA V-ROD, the W denotes the 240-mm wide rear tire. 1130 cc then 1250 cc.

VRSCX: Limited edition of 1400 with V-Rod 1250 cc engine, 240-mm wide rear tire, and official Dragster Harley-Davidson Screamin' Eagle/Vance & Hines colors.

BIG-TWIN ENGINES

Flathead: before and after 1936: 45 cu in (750 cc), 74 cu in (1200 cc), 80 cu in (1340 cc) lateral valves, engine in cast iron.

Knucklehead: 1936-1947: 61 cu in, 74 cu in balanced, engine in cast iron.

Panhead: 1948-1965: 61 cu in, 74 cu in balanced, engine in cast iron.

Shovelhead: 1966-1983 : 74 cu in, 80 cu in balanced, engine in cast iron.

Blockhead, or Evolution: 1984-1999 : 80 cu in, cylinder and cylinder heads in aluminum.

Fathead, or Twin Cam 88: 1999-2006 : 88 cu in or 1450 cc, cylinders and cylinder heads in aluminum, balanced, two camshafts driven by tensioned chains. Called Twin Cam 88, with bore x stroke of 95.3 mm x 101.6 mm mounted on the Dyna and Electra. The "Twin Cam 88 B" version had internal counter-balancers and powered the Softail from 2000 to 2006.

Twin Cam 96/96B: 2007- : 96 cu in or 1584 cc and called 1600 cc. Reutilizes the technolgy of the Twin Cam 88/88B while improving the cooling and the performance. The engine capacity is increased by maintaining the same cylinder bore and lengthening the stroke, bore x stroke = 95.3 mm x 111.1 mm.

SPORTSTER ENGINES

From 1957 to 1985, the Sportster engines resembled the Big-Twins (cylinders in cast iron, cylinder heads in aluminum). From 1986 the Sportsters are equipped with Evolution-type engines.

V-ROD ENGINES

Revolution 2002-2007: 1130 cc, V-Twin 60° angle, 4 stroke, dual overhead cams and four valves per cylinder, chain driven cams, liquid-cooled. Modern specifications for this engine, with bore x stroke of 100 mm x 72 mm, which was developed in parallel with the Twin Cam 88 then the Twin Cam 96. From November 2001 Harley-Davidson marketed this new mortorcycle called V-Rod, which was introduced in 2001 as a new model for the 2002 line.

Revolution 2008- : The same as the 1130 cc but with a capacity of 1250 cc thanks to the increased cylinder bore. Bore x stroke = 105 mm x 72 mm.

Night Rod Special VRSCDX

V-Rod VRSCAW

368 The Harley-Davidson engine often benefits from technical and aesthetic improvements, thanks to various parts conceived and built by professionals who specialize in Harley-Davidson motorcycles.

AUTHORS

Albert Saladini started his television career as a cameraman in 1976 and then became a reporter and a specialized editorial journalist. He is currently working as a journalist for France 3. Saladini is a specialist in underwater photography and an expert in cars and motorcycles. An avid Harley-Davidson enthusiast, he has virtually dedicated his life to the subject. His numerous books on the Harley-Davidson include The Harley-Davidson Story, The Loveliest Harley-Davidson and two books in collaboration with Pascal Szymezak, The American West on a Harley-Davidson and Route 66 on a Harley-Davidson. Every summer Saladini goes touring on his own Harley-Davidson, often participating to large Harley gatherings in the United States. Here, he also photographs the factories that produce the legendary bikes, and visits the customize experts.

Pascal Szymezak is a journalist who specializes in motorcycles and works for various motorcycle and motorcycle travel magazines. He has been editor-in-chief of a monthly devoted to Harley-Davidson, its history, news from America and travel in the States. He wrote The American West on a Harley-Davidson and Route 66 on a Harley-Davidson both in collaboration with Albert Saladini. He visits the States periodically for the major Harley-Davidson events. Szymezak has written many articles on the huge Harley-Davidson rallies (the Bike Week in Daytona Beach, Sturgis and Laughlin) and has interviewed the most famous Harley customizers. Every summer since 1989 he has toured the American West on his Harley-Davidson.

ACKNOWLEDGMENTS

All the historical black and white photographs come from the Harley-Davidson Motor Company Archive

Mr Willie G. Davidson ● Mrs Nancy Davidson ● Mr Bill Davidson ● Mr Steven J. Pielh ● Mr Martin Jack Rosenblum ● Mr Arlen and Mrs Cory Ness ● Mr Bob and Mrs Tracey Dron, "Harley Davidson Oakland" ● Mr Tom Worrell, "Paragon Locomotion" ● Mr Ed Kirms, "Paragon Locomotion" ● Mr Jim Betlach, "Drag Specialties" ● Mr Charlie Saloway, "Drag Specialties" ● Mr Rick Doss ● Mr Ron Simms, "Bay Area Custom Cycles" ● Mr Wyatt Fuller ● Mr Karl Smith, "Big Daddy Rat" ● Mr Claude Babot, "TechnoPlus" ● Nikon France, Mr Bruno Caron ● Agfa Gevaert S.A. France ● Mr Christian Dupont ● Laboratoire Diapro, Mr Daniel Eral ● Pix labo Photo Nice ●

*All the pictures supplied by the authors except the following:
340, 341 center and bottom, 342 top and bottom, 344, 344-345, 346 top left and right, 346 bottom, 347 top and bottom, 348 top and bottom, 350 Photo Buell*

Cover
2010 CVO FLSTSE.

Back cover
2010 CVO FXDFSE.

1906
HARLE
DAVID